Biological Monitoring of the Environment

A Manual of Methods

Biological Monitoring of the Environment

A Manual of Methods

Edited by

J. Salanki
Balaton Limnological Research Institute
Hungarian Academy of Sciences
H-8327 Tihany
Hungary

D. Jeffrey
Department of Botany
Trinity College
Dublin 2
Republic of Ireland

and

G.M. Hughes
Research Unit for Comparative Animal Respiration
University of Bristol
Brstol BS8 1UG
UK

CAB INTERNATIONAL
in association with
International Union of Biological Sciences

CAB INTERNATIONAL
Wallingford
Oxon OX10 8DE
UK

Tel: Wallingford (0491) 832111
Telex: 847964 (COMAGG G)
Telecom Gold / Dialcom: 84: CAU001
Fax: (0491) 833508

A catalogue record for this book is available from the British Library.

ISBN 0 85198 893 8

Published in association with:

The International Union of Biological Sciences (IUBS)
51 Boulevard de Montmorency
75016 Paris
France

Tel: 33 (1) 45 25 00 09
Fax: 33 (1) 45 25 20 29
Telex: c/o ICSU 645 554 F

Also published in the IUBS Methodology Series:

Research Procedure and Experimental Design for Savanna Ecology and Management
Edited by Brian Walker and J.-C. Menaut (1987)

Manual of Methods for Mountain Transect Studies
Edited by T. van der Hammen, D. Mueller-Dombois and M.A. Little (1989)

Tropical Soil Biology and Fertility: a Handbook of Methods
Edited by J.M. Anderson and J.S. Ingram (1993)
(Second Edition - published by CAB INTERNATIONAL)

Handbook of Methods for the Measurement of Work Performance, Physical Fitness and Energy Expenditure in Tropical Populations
Edited by K.J. Collins (1990)

Typeset by IUBS, Paris
Printed and bound in the UK by Information Press, Eynsham, Oxon

Table of Contents

List of Contributors

ARTEMIEVA T.I. Institute of Biology, Kazan Branch of the Russian Academy of Sciences, Kazan, Russia

BALICKA N. Department of Agricultural Microbiology, Agricultural Academy, Wroclaw, Poland

BEJCEK V. Institute of Landscape Ecology, Czech Academy of Sciences, Na sadkach 7, 37005 Ceske Budejovice, Czech Republic

BIANCHI M. C.N.R.S. Microbiologie Marine, Campus de Luminy, Marseille, France

BOHAC J. Institute of Landscape Ecology, Czech Academy of Sciences, Na sadkach 7, 37005 Ceske Budejovice, Czech Republic

BOLIS C.L. University of Milan, 9 via Balzaretti, 20133 Milan, Italy

CANCIGLIA P. University of Messina, Sicily, Italy

DOROCHOVA M.F. Moscow State University, Moscow, Russia

DOWDING P. Department of Botany, Trinity College, Dublin 2, Republic of Ireland

GAUTHIER M. I.N.S.E.R.M., U-303, Nice, France

HÄNNINEN O. Department of Physiology, University of Kuopio, SF 70211, POB 1627, Kuopio, Finland

HATAKEYAMA S. National Institute for Environmental Studies, Onogawa, Tsukuba 305, Japan

HEMMER M.J. U.S. Environmental Protection Agency, Environmental Research Laboratory, Sabine Island, Gulf Breeze, Florida 32561, U.S.A.

HUGHES G.M. Research Unit for Comparative Animal Respiration, University of Bristol, Bristol BS8 1UG, U.K.

IWAKUMA T. National Institute for Environmental Studies, Onogawa, Tsukuba 305, Japan

JEFFREY D. Department of Botany, Trinity College, Dublin 2, Republic of Ireland

KHOTKO L.I. Institute of Zoology, Byelorussian Academy of Sciences, Minsk, Byelorussia

KRIVOLUTSKY D.A. Institute of Evolutionary Animal Morphology and Ecology of the Russian Academy of Sciences, 11707 Moscow V-71, Leninsky Prospect, 33, Russia

LESNIAK A. Institute of Zoology, The Polish Academy of Sciences, ul. Wilcza 64, Warsaw, Poland

LINDSTRÖM-SEPPÄ P. Department of Physiology, University of Kuopio, SF 70211, POB 1627, Kuopio, Finland

MAUND S.J. School of Pure and Applied Biology, University of Wales, College of Cardiff, P.O. Box 915, Cardiff CF1 3TL, U.K.

McCAHON C.P. School of Pure and Applied Biology, University of Wales, College of Cardiff, P.O. Box 915, Cardiff CF1 3TL, U.K.

McKENNEY Jr, C.L. U.S. Environmental Protection Agency, Environmental Research Laboratory, Gulf Breeze, Florida 32561, U.S.A.

MIDDAUGH, D.P. U.S. Environmental Protection Agency, Environmental Research Laboratory, Sabine Island, Gulf Breeze, Florida 32561, U.S.A.

NEMCSØK J. Department of Biochemistry, Jozsef Attila University, Szeged, Hungary

PASCOE D. School of Pure and Applied Biology, University of Wales, College of Cardiff, P.O. Box 915, Cardiff CF1 3TL, U.K.

POKARZHEVSKIJ A.D. Institute of Evolutionary Animal Morphology and Ecology of the Russian Academy of Sciences, 11707 Moscow V-71, Leninsky Prospect, 33, Russia

POULTON M.J. School of Pure and Applied Biology, University of Wales, College of Cardiff, P.O. Box 915, Cardiff CF1 3TL, U.K.

ROY, S. Department of Physiology, University of Kuopio, SF 70211, POB 1627, Kuopio, Finland

RUZICKA V. Institute of Landscape Ecology, Czech Academy of Sciences, Na sadkach 7, 37005 Ceske Budejovice, Czech Republic

S.-ROZSA K. Balaton Limnological Research Institute of the Hungarian Academy of Sciences, H-8237 Tihany, Hungary

SALANKI J. Balaton Limnological Research Institute of the Hungarian Academy of Sciences, H-8237 Tihany, Hungary

SEDLACEK F. Institute of Landscape Ecology, Czech Academy of Sciences, Na sadkach 7, 37005 Ceske Budejovice, Czech Republic

SHARMA A.K. Cell and Chromosome Research, Department of Botany, University of Calcutta, Calcutta 700019, India

SHTINA E.A. Kirov Agricultural Institute, Kirov, Russia

STEGEMAN J.J. Department of Biology, Woods Hole Oceanographic Institution, Woods Hole, MA 02543, U.S.A.

TAYLOR E.J. School of Pure and Applied Biology, University of Wales, College of Cardiff, P.O. Box 915, Cardiff CF1 3TL, U.K.

USACHEV V.L. Institute of Evolutionary Animal Morphology and Ecology of the Russian Academy of Sciences, 11707 Moscow V-71, Leninsky Prospect, 33, Russia

WEGRZYN T. Department of Agricultural Microbiology, Agricultural Academy, Wroclaw, Poland

YASUNO M. National Institute for Environmental Studies, Onogawa, Tsukuba 305, Japan

Preface

The biological evaluation of the consequences of environmental pollution has brought into focus studies in both laboratory and the field and has emphasized the need for new approaches and methods. When the XXI General Assembly of the International Union of Biological Sciences (IUBS) adopted the scientific programme "*Biological Monitoring of the State of the Environment*" (Ottawa, 1982), one of the objectives was to stimulate research for improving biological monitoring of environmental pollution, to help the dissemination of the most recent and approved techniques in this field and to contribute to their introduction into practical application. During the past ten years an international cooperation has been developed on the basis of the above programme. Symposia and workshops were organized to present and discuss the use of biological monitoring methods in different countries, and the scientific results have been published in Proceedings and scientific journals.

The present manual aims to give a collection of methods tested under various conditions as being suitable for practical use in monitoring environmental pollutants or pollution. Individual papers were written mostly by members of the IUBS Commission, but also by other scientists, working in different biological disciplines. Since no rigorous editing interfered with the author's own approach and ideas, the papers are rather different in construction, style and particulars. In some sections (Microbiology and Cell Biology) the methods are condensed in one large paper by the group leader, while in others individual papers are included. Some of the papers are practical in character with only short notes on the theoretical principle but detailed descriptions for their execution. Others give a deeper scientific background and few practical instructions. Furthermore, methods for use in both field studies and/or laboratory testing are included. However, since each paper includes references recommended for further reading, it is hoped that this volume will provide a useful source of practical information for the application of bioindicator methods both in natural environments and in laboratory conditions.

The twenty-eight papers included in this volume are oriented and focused towards various areas endangered by pollution and contain different types of methods applied recently in biological indication of hazardous effects. Although only a portion of a wide and rich field can be covered and represented here, we hope this volume is a useful contribution to the efforts focused on biological monitoring of water, soil and air pollution by using living systems and living processes in predicting and monitoring harmful effects of environmental pollution. Since this programme was initiated there has been a marked increase in interest in projects of this kind. It is our hope that the work summarized here provided a part of the stimulus to the much needed expansion in this field of all types of biologist as represented by the six sections of our commission.

The Editors also wish to acknowledge the assistance and support of the IUBS Secretariat, Dr Talal Younès and Mrs Colleen Adam, in the preparation of this document.

The Editors

Microbiology

Microorganisms as Simple Ecotoxicological Bioindicators

by N. Balicka and T. Wegrzyn
Department of Agricultural Microbiology, Agricultural Academy, Wroclaw, Poland

This method, using microorganisms as ecotoxicological bioindicators, is based on their different susceptibility to xenobiotics, either singly or in complex mixtures. Differing from chemical methods, microbiological bioassays do not detect the total amount of toxic materials but their biologically active components. The approach is technologically simple, but may be calibrated against instrumental analysis.

Preparation of Microbiological Indicators

1. Several strains of bacteria and actinomyces were isolated from the soil. It is advisable to use a soil whose properties are similar to that of a polluted one. The isolation medium used contained 0.3% of yeast extract and 1% of glucose, but a soil extract of Capek medium may also be used.
2. All strains should be screened in respect to their resistance to the single xenobiotic or to the complex of xenobiotics which pollute the examined area. The most susceptible should be selected.
3. The first selection. Artificial media were used in which a range of doses of xenobiotics to be tested have been introduced.
4. The use of soil as a medium. Soil was stabilized with 1.6% of agar, supplied with yeast extract, glucose and different doses of the xenobiotics (e.g., dust emitted by copper smelter), then sterilized and poured into Petri dishes (Nikitin *et al.*, 1966).
5. The growth of bacteria and actinomyces strains on the solidified soil surface was limited by the dose of xenobiotics. The doses which reduced the growth of bacteria and actinomyces in the soil were higher than those in different artificial media.

A group of microorganisms with different susceptibility were collected, which indicated the xenobiotics level in the soil. There should be at least 30-50 strains (avoiding fungi because of their radial growth on the media surface) collected in this group in order to have multiple indications. Only one strain may be tested per plate.

Specimen Results

Determination of pollution level of the soil. We have elaborated the method for soil polluted with the industrial dust containing heavy metals (Balicka and Wegrzyn, 1979; Balicka *et al.*, 1977). Soil samples from polluted terrain were solidified with 1.6% of agar and supplied with 0.3% of yeast extract and 1% of glucose to promote microbial growth. 150 g of soil was suspended in 1 litre of medium, sterilized and poured onto plates.

The surface of the solid medium in Petri dishes was inoculated with each of the test microorganisms. The inoculation was performed as lines on the surface, 3-4 strains on one Petri dish. After incubation the number of the strains (a) normally growing; (b) with reduced growth and; (c) not growing, were noted.

This method is very easy to perform and repeat. The comparison of the growth rate for test organisms on the medium containing a standard dose of xenobiotics with that on the test soil reveal the degree of pollution (Fig. 1). It is possible to determine the extent to which pollution of the soil depends on various factors, e.g. the distance from source, wind direction, approach to cultivation. The results of determination are rather comparative. The example of estimation of the degree of pollution:

Sample A - 90% of indicator strains grow normally, 10% reduced
Sample B - 50% of indicator strains grow normally, 20% reduced
Sample C - 20% of indicator strains grow normally.

It means that sample C comes from the most polluted area.

Also to be taken into account is the xenobiotic concentration which limits growth of each strain. This provides additional information. In our experience, test results correlate with chemical analysis.

We assume that using a collection of test organisms, rather than a single strain, enhances the precision determining the extent of pollution.

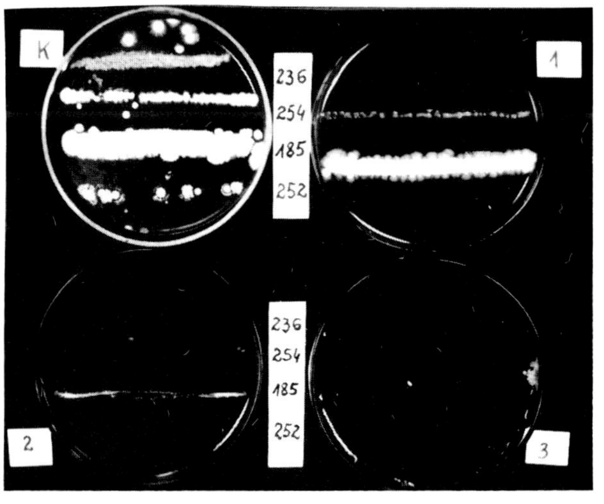

Fig. 1. The growth of indicator bacteria. Degree of soil pollution: 1 - the lowest, 2 - intermediate, 3 - the highest, K- control. 236, 254, 185, 252 - numbers of the strain.

References

Balicka, N., Wegrzyn, T., Czekanowska, E. (1977) *Acta Microb. Polon.* 20:301-308.

Balicka, N., Wegrzyn, T. (1979) Interim Symposium on Testing of Chemical Substances of Ecotoxicological Evaluation, Munich, 34-35.

Nikitin, D.I., Wasiljewa, L.W., Lochmaiseva, P. (1966) *New and Uncommon Soil Microorganisms.* Ed. Nauka, Moskwa.

Bacteriological Survey of Domestic Sewage

by Micheline Bianchi[1] and Michel Gauthier[2]
(1) C.N.R.S. Microbiologie Marine, Campus de Luminy, Marseille, France
(2) I.N.S.E.R.M., U-303, Nice, France

Purpose

The following methods are accepted procedures for the detection and enumeration of microorganisms characteristic of sewage. These include non-pathogenic indicator organisms and pathogens.

Enumeration of Fecal Coliforms

According to the O.S.I. definition, fecal coliforms are facultative aerobic Gram-negative bacteria, non-sporulated, oxidase negative, able to grow in the presence of bile salts and other surfactants, and to ferment lactose with gas production in 24 hours at 44.5°C. They produce indole in tryptophane-free peptone water at 44.5°C. They are universally used as bacterial indicators of fecal pollution in natural waters.

One of the most widely used methods for the enumeration of these indicators is the so-called "membrane filtration technique" reported here.

Specific culture medium

mFC agar (Difco): dissolve the medium in distilled water, add 10 ml of a rosolic acid solution (1% in 0.2 N NaOH) per litre of medium, mix well, cool to 50°C and immediately pour the medium into Petri dishes. This medium is to be used without further sterilization.

Procedure

Standard method

Pour the mFC medium into small petri dishes (5.5 cm diameter). Allow to solidify and stock inverted plates at 4°C in the dark. They should be used within 7 days.

Prepare a series of dilutions of the sample (as for the enumeration of total heterotrophic bacteria).

Filter 3 x 1 ml of each dilution on 3 sterile membrane filters (pore size 0.45 µm, 4.7 cm diameter) starting by the higher dilution. Wash the filter holder and the filter tower with sterile distilled water before filtering the following dilutions. Transfer the filters on mFC plates.

Place inverted plates horizontally in clean metallic boxes, seal them and incubate these tanks in a water bath at 44.5°C + 0.2°C for 24 hours.

Count typical blue colonies with a red or bronze sheen, only on membranes holding 10 to 80 colonies.

Express results as numbers of fecal coliform colonies per ml or 100 ml of sample.
For atypical colonies (faint blue): check the indole reaction and the production of β–galactosidase.

Specific references

Lechevallier, M.W., Cameron, S.C., McFeters, G.A. (1983) New medium for improved recovery of coliform bacteria from drinking water. *Appl. Environ. Microbiol.* 45:484-492.

PNUE/OMS. (1983) Determination of fecal coliforms in seawater by the membrane filtration method. Reference methods for the study of marine pollution. No. 3, Rev. 1, 26 pp.b.

Revival procedure

In some particular cases (chlorinated waters, seawater), bacterial cells are stressed and their recovery needs the use of repair or revival procedure.

Immediately after filtration, each membrane filter is placed on a non-specific medium (Trypticase Soy Agar), followed by 2 hours incubation at room temperature to effect repair. Then it is transferred to a mFC plate and incubated as described before.

Specific references

Speck, M.L., Ray, B., Read, B.R. (1975) Repair and enumeration of injured coliforms by plating procedure. *Appl. Microbiol.* 29:549-550.

Hacknet, C.R., Ray, B., Speck, M.L. (1979) Repair detection procedure for enumeration of fecal coliforms and enterococci from seafoods and marine environments. *Appl. Environ. Microbiol.* 37:947-953.

McFeters, G.A., Cameron, S.C., Le Chevallier, M.W. (1982) Influence of dilutents, media, and membrane filters on the detection of injured waterborne coliform bacteria. *Appl. Environ.* 43:97-103.

Enumeration of Enterococci

As for fecal coliforms, the enumeration of enterococci (streptococci group D) is routinely done through two methods: the membrane filtration (MF) technique and the Most Probable Number (MPN) technique using cultures in liquid medium. The MF technique, described here, is easier to develop and gives approximately the same results as the MPN method.

Specific culture medium

Medium of Isenberg *et al.* (1970) or D-coccosel medium (Biomerieux).

Standard method

Prepare dilutions of the sample and filter 1 ml of each dilution on MF as for fecal coliforms (3 MF for each dilution).

Transfer the MFs on the specific medium.

Incubate inverted plates with MF at 37°C for 24 hours.

Count typical tiny transparent colonies surrounded with a brown black halo.

Express results as numbers of enterococci colonies per ml or 100 ml of sample.

Revival procedure

Use the same technique as for fecal coliforms.

Limits

Although highly selective for Group D streptococci, the medium of Isenberg *et al.* (1970) allows the growth of several other bacteria.

Nevertheless, only *Listeria* spp. gives colonies resembling those of enterococci.

Specific reference

Isenberg, H.D., Goldberg, D., Sampson, J. (1970) Specific enumeration of fecal streptococci. *J. Appl. Bacteriol.* 20:433-436.

Enumeration of *Salmonella* spp.

Many different media and incubation conditions have been proposed for the detection and the enumeration of *salmonellas*. We describe here the procedure developed by Morinigo *et al.* (1986), grounded on the MPN technique following enrichment on a specific medium.

Media and reagents

Liquid media (tubes 10 ml):
- buffered peptone water (BPW) (Edel and Kampelmacher, 1973)
- selective Rappaport-Vassiliadis broth (Vassiliadis *et al.*, 1976) supplemented with 40 µg/ml novobiocin (RVnovo/43).

Solid media (Petri dishes 10 cm diameter):
- Xylose Lysine Desoxycholate (XLD) agar (Difco)
- Trypticase soy brilliant green (TSBG) agar (Moats, 1978)
- Trypticase soy agar (TSA)
- *Salmonella* antisera (Difco).

Procedure

Pre-enrichment

Filter 100, 10, 1 and 0.1 ml of sample through 0.45 µg pore size membrane filters (3 series of MF for each dilution).

Transfer each membrane filter into 10 ml of BPW. Incubate at 36°C for 18-24 hours.

Enrichment

From each BPW tube, inoculate 1 tube of RVnovo/43 medium in the ratio of 1:100.

Incubate at 43°C for 48-72 hours.

Isolation

Place a loopful from each of the selective broth tubes on XLD and TSBG agars.

Incubate at 36°C for 24 hours.

Isolate all typical *salmonella* colonies (XLD: red with or without black centre; TSBG: transparent with or without black centre) on TSA.

Test each isolate by standard biochemical procedures (API 20E system or conventional methods) and define its serological agglutination using commercial antisera.

Enumeration

Isolates from every tube in which *salmonellas* are biochemically and serologically confirmed are regarded as positive. The MPN of *salmonellas* per 100 ml of sample are obtained from probability tables (Standard Methods for the Examination of Waters and Wastewaters, 1981).

Specific references

Edel, W., Kampelmacher, E.H. (1973) Comparative studies on isolation methods of sublethally injured *Salmonellae* in nine European countries. *Bull. WHO.* 48:167-174.

Fricker C.R. (1987) The isolation of *Salmonellas* and campylobacters. *J. Appl. Bacteriol.* 63:99-116.

Moats, W.A. (1978) Comparison of four agar plating media with and without added novobiocin for isolation of *Salmonellae* from beef and deboned poultry meat. *Appl. Environ. Microbiol.* 36:747-751.

Morinigo, M.A., Borrego, J.J., Romero, P. (1986) Comparative study of different methods for detection and enumeration of *Salmonella* spp. in natural waters. *J. Appl. Bacteriol.* 61:169-176.

Standard Methods for the Examination of Waters and Wastewaters (1981) 15th edition, American Public Health Association,Washington DC. 830-837.

Vassiliadis, P., Petrakis, N., Papiconomou, N., Papadakis, J.,Trichopoulos, D. (1976) Nouveau procédé d'enrichissement de *Salmonella. Ann. Microbiol.* (Institut Pasteur) 127B:195-200.

Enumeration of *Vibrio cholerae* and *Vibrio parahaemolyticus*

Vibrio species of potential health importance are mostly *V. cholerae* and *V. parahaemolyticus*; even more species can be considered (Pira, 1984).

Holding conditions of samples

It is necessary to avoid keeping samples in ice or ice water; generally keep samples from direct contact with coolant. The best method is to begin enrichment procedure at sampling site.

Concentration technique of *vibrios*

To be counted in natural environments *vibrios* need to be concentrated. Two methods are available:

Membrane filter technique

The water sample (100 to 500 ml) receives Celite (filtering adjuvant prepared in stock solution and autoclaved at 121°C for 15 min) added as a final concentration of 1 g/l.

After agitation the sample(+ Celite) is filtered through a 0.45 µm filter. The filters are placed in flasks containing 50 to 100 ml of enrichment broth.

Use of gauze filtration

A tuft of gauze is attached to a wire and fixed in place in a flowing water source for up to several days, then it is transferred to enrichment broth.

Or, ten litres of water are filtered through the gauze filter. The gauze and additional water (to bring the volume to 900 ml) are transferred to enrichment broth.

Counting procedures

For both *V. cholerae* and *V. parahaemolyticus* the same methods can be used until species differentiation.

1. Counting by selective plating

Enrichment procedure

Vibrio cholerae
Two filters (see concentration technique) each are placed into 3 bottles containing 100 ml of alkaline peptone water as follows: 10 g/l peptone, 10 g/l NaCl, pH 8.5. Incubate 18-24 hours at 37°C.

Vibrio parahaemolyticus
1, 10, or 100 ml sample volumes are inoculated into the following enrichment broth: 5 g peptone; 3 g beef extract; 30 g NaCl; 0.03 g bromothymol blue; 0.001 g ethyl violet; 5

g(per litre) galactose; 9.0 pH. Incubate 24 hours at 37°C.

Counting procedure

Culture medium: TCBS (thiosulfate citrate bile salts) (BBL). Mix 86 g of dehydrated medium and boil it for one minute. Cool to 50°C and pour into plates. This medium does not require autoclaving.

Procedure

After enrichment in alkaline peptone water, streak onto TCBS and incubate at 37°C for 24 hours.

If no colony appears on TCBS, keep the enrichment broth one week at 100°C and streak again on TCBS plates.

Isolated colonies are screened for oxidase reaction; oxidase positive colonies being purified on tryptic soy agar (Difco).

Species differentiation

Prepare: - salt-free gelatine neopeptone agar
- gelatine agar with 1% NaCl

Inoculate a quarter plate of each agar with isolated colony.

Two 6 mm filter disks containing 10 to 150 µg of 0/129 vibriostatic compound are placed on each of the quadrants of the gelatine agar plate.

After overnight incubation at 37°C score: ability to grow in absence of NaCl; gelatinase production; 0/129 sensibility.

Presumed *V. cholerae* strains:
Yellow colonies on TCBS
Growth without NaCl
Sensitive to 10 µg 0/129

Presumed *V. parahaemolyticus* strains:
Green colonies on TCBS
No growth without NaCl
Sensitive to 150 µg 0/129

Limits

These methods are easy to use for routine laboratory practice, but precise taxonomic identification needs more tests. The final step of such a study is to confirm that isolates are really pathogenic, and needs substantial effort.

Specific references

Kaper, J., Mockman, H., Colwell, R.R., Joseph, S.W. (1979) Ecology, serology and enterotoxin production of *Vibrio cholerae* in Chesapeake Bay. *Appl. Environ. Microbiol.* 37:91-103.

Kobayasishi, T., Enomoto, M.S., Sakazaki, R., Kuwahara, S. (1963) A new selective isolation medium for pathogenic *vibrios* TCBS agar. *Jap. J. Bact.* 18:387-391.

Monsur, K.A. (1963) Bacteriological diagnosis of cholera under field conditions. *Bull. WHO.* 28:387-389.

Pira, W.M. (1984) *Vibrios in the environment.* R.R. Colwell (ed.) Wiley and Son, N.Y., 251-268.

2. Direct counts by fluorescent antibody staining method

Procedure

1. Collect water samples in sterile containers employing aseptic techniques. Add filter-sterilized and autoclaved solution of yeast extract to obtain a final concentration of 0.025%. It is recommended that the concentration of yeast extract be prepared so that 1.0 ml of 0.25% yeast extract solution can be added to a 10 ml sample.
2. Incubate for 20 min at a temperature appropriate to the environment from which the sample was collected.
3. Prepare an 0.2% solution of nalidixic acid (Sigma Chemical) in 0.05 N NaOH, filter sterilize, add 0.1 to 10 ml of sample.
4. Incubate at appropriate environmental temperature for 12 hours.
5. Filter the water sample through an 0.2 µm x 25 mm polycarbonate membrane filter (Nuclepore), mounted on a 25 mm glass filtration unit. Filters are prestained overnight in 0.2% irgalan black (Ciba-Geigy) dissolved in 2% acetic acid and rinsed several times in distilled water before use, or use black Nuclepore filters. Place filters in a 30 x 10 mm Petri dish.
6. Fix cells to filters by placing Petri dishes on the surface of a test tube rack partially submerged in a 55°C water bath. Incubate for 20 min.
7. Prepare a 1:20 dilution of rhodamine isothiocyanate-bovine serum albumin. Place one drop on the surface of the membrane, and place a coverslip on the top of the membrane to spread the reagent across the filter.
8. Incubate at 35-37°C for 30 min in a dark moist chamber.
9. Rinse filters three times with phosphate-buffered saline (PBS: 8.5 g NaCl; 1.5 g Na_2HPO_4; pH 7.3) and allow the filters to soak for 10 min in the Petri dish. Remove PBS and air dry the filters. Care must be taken to ensure that the top of the filter always remains upright throughout the procedure.
10. Add one drop of *Vibrio cholerae* 01 antiserum and place a coverslip on the top of the drop.
11. Repeat steps 8 and 9.
12. Add one drop of fluorescein isothiocyanate anti-rabbit goat serum (FITC, BBL). Place coverslip on top.
13. Repeat steps 8 and 9.
14. Mount each filter on a glass slide with a coverslip using FA mounting fluid, pH 9 (Difco).

Examine with a microscope equipped with a neofluor objective, IVFL epifluorescence condenser, 100 W halogen lamp, BP 450-490 band pass filter, FT 510 beam splitter and 520 barrier filter, or equivalent. Slides should be examined as soon after preparation as possible to avoid fading of the intensity of the fluorescence. Slides should be stored in a dark box until used. The *V. cholerae* 01 cells that are counted are those that appear as long and/or fat rods possessing a peripheral green band just below the cell wall. Calculate total viable cells per ml of sample, according to the size of the ocular reticule in the microscope.

Limits

The major problem of this method is the specificity of the antiserum used, and possible cross-reactivity with other bacterial species must be controlled. Each new batch of antiserum must be tested.

Direct microscopic counts can be difficult in highly turbid waters because of the number of particles in each field.

Specific reference

Brayton, P.R., Colwell, R.R. (1987) Fluorescent antibody staining methods for enumeration of viable environmental *Vibrio cholerae* 01. *J. Microbiol. Methods* 6:309-314.

Cell Biology

Cell Biology Techniques for the Detection of Toxic and Mutagenic Substances

by A.K. Sharma
Cell and Chromosome Research, Department of Botany, University of Calcutta, Calcutta 700019, India

Introduction

The methods based on cell biology cover both indication and monitoring of the state of the environment. The tests principally relate to an analysis of toxicity, mutagenicity and carcinogenicity of the materials prevalent in the environment (Sharma and Sharma, 1960; Oftedal, 1979; Sharma, 1981; Sugimura *et al.*, 1982; Fishbein, 1984; Sobels, 1984). Different biological systems have been widely employed as mutagen indicators for a long time (Zimmerman and Taylor-Mayer, 1985). For *in situ* monitoring, plant systems can serve as ideal materials for monitoring of polluted air, water, industrial chemicals, agricultural chemicals and ionizing radiation (Sharma, 1986). In the present manual, a series of tests based on genetic assay of both lower and higher organisms as indicators have been outlined.

The cell biological methods for monitoring and indication of damage are principally concerned with genetic systems. However, in addition to the nucleus, other organelles are also affected by environmental pollutants. In order to assess the damage to all the organelles, ultrastructural analysis in addition to fluorescent microscopy should be adopted. One of the limitations of the cell biological techniques is the fact that several compounds present in air, soil and water can often lead to similar types of damage. The monitoring of the damage caused by environmental pollutants can be precisely carried out following the techniques mentioned above. But in order to associate the effects with the precise source of the damage it would be desirable to combine histochemical techniques along with cell biological methods.

References

Fishbein, L. (1984) *Genetics: New Frontiers*, Chopra, V.L., Joshi B.C., Sharma, R.P., Pran, H.L. (eds). Oxford and IBH Co., New Delhi. 3:42.

Oftedal, P. (1979) In: *Genetic Damage in Plants caused by Environmental Agents*, Berg. K. (ed.). Academic Press, N.Y.

Sharma, A.K. (1981) Impact of the Development of Science and Technology on Environment. Presidential Address. 69th Session. Ind. Sci. Cong. Assoc.

Sharma, A.K. (1986) Cell biological methods as indicators of environment. In: *Bioindicators*, Salanki, J. (ed.). IUBS Monogr. Series, IRL Press, U.K. Vol°1:35.

Sharma, A.K., Sharma, A. (1960) *Int. Rev. Cytol.* 10:101.

Sobels, F.H. (1984) *Genetics: New Frontiers*, Chopra, V.L., Joshi B.C., Sharma, R.P., Pran, H.L. (eds). Oxford and IBH Co., New Delhi. 3: 63.

Sugimura, T., Konde, S., Takebe, H. (1982) *Environmental Mutagens and Carcinogens.* University of Tokyo Press, Tokyo.

Zimmerman, F.K., Taylor-Mayer, R.E. (1985) *Mutagenicity Testing in Environmental Pollution Control.* Ellis Horwood, Ltd, Chichester, U.K.

PART I

Salmonella as a Test System

The *Salmonella* back mutation system (Ames' Test)

The series of *Salmonella typhimurium* strains developed by Bruce Ames for use as testers for back mutation at the histidine locus of TA strains, constitute the most widely used and one of the most useful systems for identifying genotoxic genes (Ames *et al.*, 1973). The principle is to apply the substance to be tested in suitable concentrations and under suitable conditions to several strains of *Salmonella* carrying different molecularly defined mutations in the histidine locus. Back mutations are identified, and from the type of strain showing back mutations and from the conditions necessary to elicit this response, it is possible to characterize the type of mechanism underlying the genotoxic action. Deeper understanding of the process of carcinogenesis has made it clear that carcinogenesis may be induced without collateral mutagenesis. The system does not easily lend itself to quantitative analyses of mutagenic potency, but does provide a sensitive and widely used qualitative probe for monitoring genotoxic activities in single chemicals or mixtures.

The strains are characterized by their specific and different mutations - base substitution or frameshift - in the histidine locus. In addition, all strains have genetic constitutions which make them more susceptible to penetration and attack by genotoxic agents. Mutation rfa makes the cell wall more penetrable to substances in the medium. Mutation uvrB blocks some of the normal repair functions which would ordinarily reduce the yield of back mutations. The plasmid pKM101 enhances a repair mechanism which is error-prone, thus introducing an amplification system which again increases the sensitivity of the strains to a genotoxic insult. Additional strains have been constructed for specific purposes: TA97 is similar to TA1537 but more sensitive to certain chemicals, the TA102 is sensitive to some agents poorly detected by the other strains (UV, H_2O_2, formaldehyde).

Technique

The test bacteria are exposed to the test substance in a metabolically active condition. For this reason, the bacteria are seeded in an agar overlay which contains trace amounts of histidine, so that all cells are able to go through a small number of divisions. However, continued growth is dependent on a reverse mutation taking place, leading to histidine independence. Comparison of the number of revertant clones per plate in test plates and in controls will then provide a measure of the mutagenic response. Because some of the strains are less stable, it is important to have the controls treated in a manner which links a response in the test plates unequivocally with the exposure to the test substance. This implies that the controls must be exposed to all components of the test treatment except exposure to the test substance itself.

Ames' tester strains can be obtained from Professor Bruce N. Ames, Department of Biochemistry, University of California at Berkeley, CA 94720, U.S.A.

Any response to the test treatment is identified as dose dependent. In the context of single identifiable substances, this means exposure to a series of concentrations ranging from zero to slightly toxic levels.

In higher organisms, many substances acquire genotoxic properties only in the course of their metabolic degradation. These complexities may be simulated in the *in vitro* tests by adding a cell extract containing the relevant enzyme complexes (P450) together with the necessary energy sources, constituting the so-called S9 mix. The cell extracts usually consist of the microsomal fraction of mammalian (rat) liver from animals that have been pretreated with a substance (e.g., Aroclor 1254) which induces/stimulates the production of these detoxifying enzymes.

S9 mix. The 9000 g supernatant of adult rat liver homogenate from Aroclor 1254 induced animals should contain: NADP (sodium salt) - 4 µmol/ml, D-glucose-6-phosphate - 5µ mol/ml, $MgCl_2$ - 8 µmol/ml, KCl - 33 µmol/ml, sodium phosphate buffer, pH 7.4 - 100 µmol/ml and S9 fraction from homogenate - 100 µl/ml.

If the test sample shows increased reverse mutation frequency to three different concentrations, and the highest response is 2-3 times the control value, the response may be considered positive. Single, marginal or sporadic increases should not be given much weight. Similar responses from similar strains (frame-shift, or base substitution) strengthen the evidence for a positive response.

This method has been further modified by TLC/Salmonella technique involving the separation of complex samples on this layer chromatography and registering of mutagenic effect directly on the plates by the bacterium (Holler *et al.*, 1985)

TLC/Salmonella Protocol

1. Select two strains TA98 and TA100 of *Salmonella typhimurium* and perform mutagenicity test with TA98 and TA100 with or without S9 mix prepared from rat liver.
2. Develop TLC plates by placing in the bottom of petri dishes. Add 30 cc minimal agar carefully to each dish.
3. Pour liquid agar in a part of the bottom of the dish not covered by the TLC plate. Distribute the agar slowly over the surface of the plate and keep the agar layer very thin.
4. After one to two hours, take 0.3 cc of the tester strain with or without S9 mix (liver homogenate) and mix with 6 cc of top agar. Pour it on the minimal agar.
5. After 48 to 72 hours at 37°C observe histidine revertants in area containing mutagens.

Experiments can be performed with or without the addition of Aroclor-induced liver homogenate. The outdoor mutagenicity is expressed as revertants per 30 m^3 of air and indoor mutagenicity is used as rev/m^3 both calculated from the slope of the linear part of the dose-effect curve. EMS and DMSO can be used as positive control.

References

Ames, B.N., Durston, W.D., Yamasaki, E., Lee, F.D. (1973) *Proc. Natl. Acad. Sci. USA.* 70: 2281.

Brusick, B. (1979) *Principles of Genetic Toxicology.* Plenum Press, N.Y. and London.

Long-Term and Short-Term Assays for Carcinogens. (1986) IARC Scientific Publication No. 83.

Holler, M., Bjorseth, A., Honk, V.S. (1985) In: *Mutagenicity Testing in Environmental Pollution Control.* Zimmerman, F.K., Taylor-Mayer, R.E. (eds). Ellis Horwood Ltd, Chichester, U.K.

Yeast as a Test System

Strains of yeast have been utilized as indicators of reverse mutations, gene conversion, aneuploidy and other chromosome damage (Zimmerman *et al.*, 1975; Nestman, 1985). Growing cells of *Saccharomyces cerevisiae* produce the enzymes necessary to activate many promutagens, which would not otherwise be mutagenic in *in vitro* assay without addition of S9 mix (Callen *et al.*, 1980) as noted in Ames' test.

Yeast-Agar Overlay Protocol (Parry 1985)

1. Suspend stationary-phase cells of an appropriate tester strain in saline, 107 cells cc.
2. Take the test extract up to 4% and mix with yeast minimal medium* kept at 45°C with requisite number of yeast cells. Normally $35x10^3$ cells per cc and $35x10^2$ cells per cc are adequate for the detection of histidine - independent and tryptophan - independent prototrophs respectively. If needed, an exogenous S9 mix may be added.

* Supplement the minimal media with 20 μg cc^{-1} of tryptophan and 0.1 μg cc^{-1} of histidine for detection of respective prototrophs as used. This allows at least 3 cell divisions during incubation along with the tissue extract.

3. Keep the plates in the dark for 9 days at 28°C and count the frequency of prototrophic colonies.

References

Callen, D.F., Wolf, C.R., Philpot, R.M. (1980) *Mutat. Res.* 77:85.

Nestman, E.R. (1985) In: *Mutagenicity Testing in Environmental Pollution Control.* Zimmerman, F.K., Taylor-Mayer, R.E. (eds) Ellis Horwood Ltd, Chichester, U.K.

Parry, J.M. (1985) In: *Mutagenicity Testing in Environmental Pollution Control.* Zimmerman, F.K., Taylor-Mayer, R.E. (eds) Ellis Horwood Ltd, Chichester, U.K.

Zimmerman, F.K., Kern, R., Rosenberger, H. (1975) *Mutat. Res.* 28:31.

Maize as a Test System

Higher plants have been utilized for genetic monitoring of environmental mutagens because of the complexity of their genomes, comparable to mammals, and their capacity to activate a promutagen to a mutagen. *Zea mays* is a very good bioindicator because of the convenience of detecting mutation in the specific gene locus of identified chromosomes. The two tests which are widely employed involve chromosome 9.

I. The Waxy Locus Pollen Test

It is located in the short arm of chromosome 9 and is responsible for the synthesis of carbohydrate amylose. The starch of individuals carrying the dominant allele *Wx* is a mixture of amylose and amylopectin, whereas the double recessive individuals do not synthesise amylose (Sprague *et al.*, 1943; Plewa 1981, 1985). Iodine test leads to blue/black coloration of amylose and in its absence a red or tan colour is produced. As the pollen grain is haploid, containing either *Wx* or *wx*, the detection of colour following the pollen test is simple.

Protocol (Plewa, 1985)

1. Take an Early-synthetic type of maize and allow it to grow till anthesis in an area containing the mutagenic pollutants.
2. Harvest each tassel from double recessive *sx*-C/*wx*-C and label.
3. Dehydrate to 70% ethanol for 48 hours.
4. Shake the selected tassels in pure ethanol and remove a few unopened flowers.
5. Dissect the anthers from the floret and place in a stainless steel cup of microhomogenizer.
6. Prepare: Stock A - Iodine solution with 25 cc of water, 100 cc of KI and 95 mg of Iodine. Stock B - gelatine solution with 25 cc of hot water and 15 g of gelatin. Before using mix A and B in equal proportion.
7. Take 0.6 cc of this mixture and add in the cup. Add a drop of Tween-80.
8. Homogenize the minced anthers for 30 seconds.
9. Filter the homogenate through cheese cloth on a microscope slide and put the cover slip on the suspension.
10. When the suspension solidifies, examine the slide under the microscope. The black grains would indicate the revertants and the percentage of mutation is worked out.

II. Yg-2 locus test

This locus of short arm of chromosome 9 leads to pale yellow green and slowly maturing leaves in homozygous recessive plants *Yg*-2/*yg*-2. The dominant *Yg*-2 leads to full green colour. In the heterozygous state *Yg*-2/*yg*-2 shows normal green leaves. In a heterozygous individual, mutation of the dominant allele leads to deficiency in the production of chlorophyll in the leaf sector. If a leaf primordial cell is involved, the leaf ultimately leads to yellow green sectors in the green leaf (Dowd and Plewa, 1983; Plewa, 1985).

Protocol

1. Choose heterozygous *Yg*-2/*yg*-2 seeds from Early-Early synthetic type.
2. Expose the seeds under aeration to specific environmental pollutants for which this bioindicator test is applied by immersion.
3. Germinate the seeds and grow the plant in growth chambers at 28°C till the emergence of 4 to 5 leaves.
4. Dissect the mature leaves with ligule and observe under the ultra-violet light. Yellow-green sectors can be clearly seen under magnifying lens.

III. Maize micronuclei test (Wagner and Plewa, 1983)

In the Early-Early synthetic type, micronuclei tests in root tips give an indication of the chromosome aberrations produced as a result of the physical and chemical pollutants in the environment. This method can be utilized as an effective indicator and monitor.

Protocol

1. Collect young tips from germinating seeds of plants grown in a specific environment which is to be monitored.

2. Fix for 24 hours in ethanol-acetic acid mixture 3:1. Rinse in demineralized water.
3. Hydrolyse in 1N HCl at 60°C for 7.5 minutes.
4. Rinse the root tips again and place in Feulgen solution for 1 hour.
5. Treat the root tips with pectinase solution for softening and clearing.
6. Smear the root tips in 45% acetic acid.
7. Count the number of interphase cells under microscope. At least 1000 interphase cells per slide should be scored.

It is a convenient and rapid assay method for monitoring acute or chronic exposure of genes and shows direct correlation with the degree of pollutants or mutagens.

References

Dowd, P.A., Plewa, M.J. (1983) *Maize Gen. Crop News Lett.* 57:144

Plewa, M.J. (1981) *Mutat. Res.* 78.

Plewa, M.J. (1985) In: *Mutagenicity Testing in Environmental Pollution Control.* Zimmerman, F.K., Taylor-Mayer, R.E. (eds). Ellis Horwood Ltd, Chichester, U.K.

Sprague, G.F., Brimhall, B., Hison, R.M. (1943) *J. Am. Sr. Agron.* 35:817.

Wagner, E.D., Plewa, M.J. (1983) *Maize Gen. Crop News Lett.* 57:141.

Tradescantia as a Test System

I. Tradescantia staminal hair method

A change in colour in highly mutagen sensitive staminal hairs has been utilized to monitor air pollutants as well as physical and chemical mutagens (Schairer *et al.*, 1978; Fugii, 1982). The mutational events cause a change in colour from blue to pink. The technique simply involves the dissection of the stamen and noting the colour of staminal hairs on a mounted slide under the microscope.

II. Tradescantia microspore method

Conger (1953) utilized the microspore of Tradescantia as a biological dosimeter to measure the extent of radiation as absorbed by the plant when flown through a radioactive cloud. In general, the microspores of Tradescantia serve as ideal materials for scoring the effects of pollutants as manifested by chromosomal aberrations during mitosis of pollen grains.

Protocol (Conger, 1953)

1. Pick up pollens with a brush from anthers of plants growing in specific polluted environments.
2. Heat 12 g of lactone and 1.5 g of agar with 100 ml distilled water till agar dissolves. Add 0.01 g of colchicine when the temperature cools down to 80-60°C. Keep at 60-70°C.
3. Coat slides with a thin layer of egg white. Dip in the medium at 60°C until it warms up. Withdraw the slide, drain off the medium, and wipe the back with a piece of clean cloth.
4. Dust a thin layer of pollen on the medium after it has set. Immediately place the slide in a moist growing box which is a horizontal glass staining dish lined with damp blotting paper on the two sides and top, at 20-25°C. Observe at intervals till the optimum period for maximum division is found.

5. Add a drop of 1% acetic-carmine solution to the pollen tube. Squash under a cover glass and observe.

In addition to maize and *Tradescantia*, barley (*Vicia faba*) as well as species of *Arabidopsis* offer good systems for environmental monitoring. In species of *Arabidopsis*, the advantage is that because the entire life cycle is of short duration, the effect of the pollutant can be measured in all phases of growth.

References

Conger, A.D. (1953) *Stain Tech.* 28:289.

Fujii, T. (1982) *Environmental Mutagens and Carcinogens.* Univ. Tokyo Press, Alan R. Liss., N.Y.

Schairer, L.A., Van't Hof. J., Hayes, C.G., Burton, R.M., de Serres, F.J. (1978) *Environmental Health Perspectives* 27:51.

Allium as a Test System

The bulbs of onion provide an ideal monitoring system to detect chromosomal damage caused by environmental agents. The rapid root growth in succession, comparatively large chromosomes and convenience in fixation and staining make this material ideally suited as a test system (Levan, 1949).

Protocol

1. Select healthy bulbs of *Allium cepa* from affected environment and grow on the top of a jar containing Knop's solution. Alternatively, the pollutant may be added to the culture solution. Root tips are removed and scored successively.
2. Keep the jar in a temperature of 25-30°C.
3. Excise young healthy root tips and fix in acetic-ethanol mixture 1:2 for 30 minutes.
4. Transfer to 45% acetic acid for 15 minutes.
5. Prepare a mixture of 2% acetic-orcein in 45% acetic acid and 1N HCl mixture in the proportion of 9:1.
6. Slightly warm the root tips on flame for 4 to 5 seconds for staining, separation of middle lamella and clearing.
7. Keep in the stain mixture for 1 hour.
8. Mount in 1% acetic orcein or 45% acetic acid.
9. Observe the number of chromosome fragments in metaphase and anaphase stages. The frequency of breaks gives an indication of the extent of chromosome damage caused by the environmental pollutants.

Orcein staining may be applied carefully as overheating may lead to chromosome breaks (Levan, 1949; Sharma and Sharma, 1980). If necessary, in place of orcein, Feulgen staining can also be applied.

References

Levan, A. (1949) Proc. Int. Congr. Genet. Stockholm. *Hereditas Suppl.* 325:337.

Sharma, A.K., Sharma, A. (1980) *Chromosome Techniques - Theory and Practice.* Third Edition. Butterworths, London.

PART II

Drosophila as a Test System

The fruit fly *Drosophila melanogaster* is a very good indicator animal for *in vivo* testing of genetic damage at the chromosome to gene level (Hsu, 1982). It has a great potential for use in monitoring biologically active substances in the environment. By the use of the wild type as the exposed species, the response becomes more realistic than by the use of genetically manipulated species which may be highly efficient registrars of changes, but where the implications for other organisms or the ecosystem may be difficult to quantify. In that sense *Drosophila* has an important role in the array of test systems which may be used for monitoring the state of the environment.

If a continuous monitoring of, for example, atmospheric pollution in a working environment is the object of study, a set of population cages kept at strategic points for long-term exposure may be the most efficient method. If, alternatively, particulate matter filtered from large air volumes, or soil or water from a particular case of pollution is the problem, suitable exposures of conventional vial or bottle cultures over a single generation may be a better method.

Population cages are to be exposed mainly to gaseous substances for periods of several generations, at ambient temperature. If the temperature is about 25°C, the generation time would be 10-11 days, and a bi-weekly turnover of breeding-cups may be in order. Individual exposures may be either by feeding or injection of larvae or imagoes, males or females. Feeding or injecting larvae may lead to exposure of early stages of germ cell maturation. Feeding or injecting imagoes leads to exposure of all stages of maturing germ cells.

Dominant lethal test

Dominant lethals, with their non-accumulative nature, appear to be more suitable for the tracking of the immediate effects of a genotoxic substance. A dominant lethal may act at any stage prior to reproduction in the F1, and does, genetically speaking, include F1 sterility. Usually, however, one concentrates on those mutations that prevent the embryo from hatching from the egg. The test is performed by mating an exposed male or female to an unexposed individual of the opposite sex. Eggs are collected over a suitable period (for example, overnight), and counted. Those eggs which are unhatched are counted 24 hours later. In order to avoid unfertilized eggs confusing the issue, exposed virgin females are mated to 2-3 unexposed males, and conversely, exposed males are given at least 24 hours to mate with one unexposed virgin female. Concurrent controls are similarly treated.

It is simplest to collect eggs from a batch of females using filter paper moistened with acetic/alcohol (1% + 1%).

Testing for ***recessive sex-linked lethals*** (sll) implicates only offspring that survive and are fertile to produce an F2 generation. The principle of the test is to isolate single X-chromosomes and register if this chromosome allows the imago to survive when hemizygous. The test is excellently suited for quantification of the genotoxicity and its invention by H.J. Müller enabled him in 1927 to demonstrate the mutagenic effect of X-rays.

Since the mutations are by definition recessive, the test may be used to reveal damage which has accumulated over several generations in chromosomes inherited through the female line and thus not subject to selection in the hemizygous condition.

Müller 5 test

Individual X-chromosomes are sampled either in exposed virgin females or in exposed males. By mating to Basc (or M5 as they are usually called) or Binsey flies of the opposite sex, F1 females with one M5 and one exposed X are produced. Each such heterozygous female represents one exposed X to be tested. When she is mated, usually to M5 males, she will produce M5 sons, but if a lethal has been induced in the other X-chromosome, no wild type sons. The measurement of the effect thus emerges from the number of F1 females that do not produce wild type sons, relative to the total number of fertile F1 females tested. The sll test is thus characterized by the large number of F2s that need to be raised, but also by the simple and rapid scoring possible since wild type males have large round red eyes, while M5 males have narrow pale apricot coloured eyes. The difference can be seen without magnification. All cultures with red eyed males are discarded.

The other sex-linked recessive lethal test is ClB where the male fly is generally exposed for studying sex-linked recessive lethal in order to avoid the possibility of pre-existing lethal mutation in the X-chromosome of female.

As indicators of point mutation, after *in situ* exposure of the male fly to the specific environment, the fly is to be bred with a specific tester strain.

ClB test

Protocol

Allow the treated fly (after usual exposure) to grow in the culture medium specially designed for *Drosophila.*

The medium may be prepared in the following way:

1. Add 1.5 g agar to 75 cc of water, boil to dissolve and keep overnight. To the medium add 10 g of commercial Treacle 13.5 cc and a trace of preservation Nipagin or Moldex. Mix thoroughly and add a few drops of yeast in it. Cool, transfer a small quantity of the food to small culture bottles 3" x 1" and culture the fly.
2. Take ClB stock of female fly (Bar-eyed, Crossover inhibiting and containing a recessive lethal in one X-chromosome) and mate with treated male fly in the medium specified above.
3. After 48 hours, transfer the flies to bottles prepared by placing a quantity of food about 1 inch deep together with a piece of folded sterilized crepe paper and plug the bottle with wool.
4. After the eggs are laid, remove the parents to a different bottle.

5. After the larvae come out, give the cultures another yeasting.
6. Sort out the Bar-eyed females. In such females, one X contains the pre-existing lethal gene and the other X is from the treated male.
7. Mate these females with normal males following the above procedure.
8. Culture the larvae following the procedure as outlined in steps 4 and 5.
9. Score the flies for maleness. As Y is genetically inactive in relation to this factor and males would have one of the "X"s of either ClB or treated X, there will be no male fly if there is recessive lethal mutation in the treated X.

These methods are widely used as very powerful indicator tests for environmental mutagenesis.

There are many other aspects of *Drosophila* genetics which may be utilized for monitoring the environment.

First, *non-disjunction* or maldistribution of chromosomes should be mentioned, primarily because the character and mechanism of induction of this type of damage are largely different from that discussed above, which concerns damage to DNA. Non-disjunction is caused primarily by damage to the chromosome's centromere, or by disturbance of the spindle apparatus which leads to disturbance of the orderly distribution of chromosomes to daughter cells.

The test requires that the aneuploid offspring survives, and therefore is limited to the X-chromosome or the small fourth chromosome. In practice only the X-chromosome is used. In order to distinguish the offspring resulting from non-disjunction, one set of chromosomes must carry markers.

If an exposed female produces a non-disjunction egg, it will be either XX or nullo-X. The egg will be fertilized by an X- or Y-carrying sperm, which in the first case will lead to an XXX individual which dies, or an XXY female which shows the parental female phenotype. If the male X carries a dominant marker, these exceptional females can be distinguished. The nullo-X egg will be fertilized either by a male parental X which again will lead to an exceptional individual, since all other males are of the exposed mother's genotype, or a nullo-X egg fertilized by a Y carrying sperm which will not survive.

For non-disjunction in the male, the exposed male carries an X with a dominant marker, and a marker on its Y. Non-disjunction sperm may then be picked up in wild type females. Non-disjunction in the non-exposed sex may confuse the issue somewhat, but this can be resolved by utilizing the additional dominant marker on the male (e.g. Bar).

Finally, tests for *specific locus mutations* should be mentioned. The method consists simply in crossing the exposed individual to a tester which is homozygous for a maximum number of recessive markers. If an exposed chromosome has a dominant mutation, or a recessive mutation at one of the marker loci, or a non-lethal deficiency uncovering one of the marker loci, this can be recognized in the individual carrying that exposed chromosome.

References

Ashburner, M. (1986) (ed.) *The Genetics and Biology of Drosophila.* 1-3. Academic Press, N.Y.

Hsu, T.C. (1982) *Cytogenetic Assays of Environmental Mutagens.* Allanheld Osman Publ., N.Y.
Roberts, D.B. (1986) *Drosophila: a Practical Approach.* IRL Press.
Demerec, M. (1950) (ed.) *Biology of Drosophila.* John Wiley, USA.

Fish as a Test System

This system has been utilized for monitoring of the aqueous environment and chromosome damage caused thereby. The chromosome damage is principally measured through chromosome breaks, bridges and other aberrations. The organs normally utilized for tests in fish are the gill cells in addition to kidney and intestine. Laboratory fish such as *Ubra pygmaea* (mudminnow) can be exposed for different periods in various aqueous environments (Prein *et al.*, 1978).

The two key steps in most vertebrate test systems are: treatment with colchicine to secure a large number of metaphases for scoring and fixation on a hypotonic medium.

Protocol (Kligerman, 1982)

1. Select healthy fishes from specific environments under analysis.
2. Approximately 7 hours prior to sacrifice, inject fishes with 25 μg/g colcemid or colchicine and return to well-aerated aquaria.
3. Sacrifice fish by decapitation. Dissect out gills, kidneys, intestines and place in 10 times its volume of 0.4% KCl hypotonic solution for 30 min.
4. Remove tissue and place in 3:1 ethanol or methanol/acetic acid fixative for 30 min. Remove fixative and replace with fresh fixative for 2 hours to overnight.
5. Remove tissue, touch briefly to absorbent paper. Immediately place in the depression of a concave slide. Add 2-3 drops of 50% acetic acid.
6. Mince the tissue with forceps to form cell suspension and return unsuspended tissue to fixative.
7. Withdraw 30 μl of suspension with micropipette or microhaematocrit capillary tube and expel it into a clean glass slide heated to 40-60°C on a slide warmer.
8. After 5 seconds, withdraw the suspension from the heated slide back into the capillary tube or pipette leaving behind a ring of dried cellular material 1-2 cm in diameter.
9. Repeat steps 7 and 8 producing 2 or 3 rings per slide.
10. Stain slides in 4% Giemsa in 0.01 M Phosphate buffer (pH 7) for 10 min.
11. Rinse slides in distilled water, air dry, and place in xylene for at least 10 min.
12. Mount with suitable mounting media using glass coverslips.

References

Kligerman, A.D. (1982) In: *Cytogenetic Assays of Environmental Mutagens.* Hsu, T.C. (ed.). Allanheld. Osmun Publishers, N.Y.
Prein, A.E., Thie, G.M., Alink, G.M., Poels, C.L.M., Koeman, J.A. (1978) *Sci. Total Environ.* 9:287.

Mice as a Test System

The mouse as a test system can be regarded as very crucial for monitoring of mutagenic potential of any constituent of the environment.The indications of toxicity are reflected in the production of lethals or chromosome aberrations as in *Drosophila.* The death of the embryo which is the indication of lethality can be judged by scoring of dead implants (*vide* Soares, 1972; Hsu, 1982; Manna and Ghosh, 1983). Chromosome aberrations can be studied either in the bone marrow or in leucocytes.

I. Dominant lethal test

Protocol

1. Take ten old mice from a polluted environment.
2. Mate each male separately with 7 sets of virgin females, one set each week for 7 consecutive weeks.
3. Determine successful mating by the appearance of vaginal plug in the mating females.
4. Sacrifice two pregnant females on the 16th day of gestation.
5. Expose the viscera and clear the two horns of the uterus.
6. Determine the number of live implants by the conspicuous swelling at the place of implant and the dead implants by the scar mark.
7. Keep records separately for 7 weeks.
8. Sacrifice one or two mating females at the end of the weaning period after parturition, each week.
9. Expose and remove carefully the uterus and put it in a petri dish containing normal saline.
10. Slit open each horn and examine the wall of the lumen under a dissecting binocular microscope and count the number of brownish scars surrounded by significant amounts of vascular tissue representing living implantation and the light brown or whitish scar marks with less surrounding vascular tissue, representing dead implantation.
11. Combine the data of different mothers of a week and find out the number of living and dead implants per mother each week.
12. Use the following formula to determine the post implantation lethality %:

$$= 1 - \frac{\text{n}^\circ \text{ of live implants/total imp. in exp gr x 100}}{\text{n}^\circ \text{ live implant/total imp. in control gr}}$$

and

$$\text{Mutagenic index} = \frac{\text{n}^\circ \text{ of dead implants x 100}}{\text{total number of implants}}$$

13. Subject the data of different weeks to a statistical significance test and find out if there is stage sensitivity according to the timing of spermatogenes.
14. If data are insufficient, pool the data of all weeks and analyse statistically.

II. Chromosome aberration test

Protocol (*vide* Hsu, 1982; Sharma and Sharma, 1980)

1. Inject 0.5 cc of 0.025% (w/v) colchicine solution intraperitoneally in the exposed animal. Leave for 1 hour or more. Excessive toxicity should be avoided. Prolonged keeping may result in polyploidy.
2. Kill. Remove femurs and cut off epiphyses. Wash out the marrow into a small phial with warm 0.12% aqueous solution of sodium citrate, using a hypodermic syringe with a fine needle.
3. Gently aspirate the marrow in and out of the syringe till it breaks up into a fine suspension. Keep the phial with the suspension in a water bath at 37°C for 20 minutes.

4. Filter by centrifuging the suspension through nylon bolting cloth in a bacterial infiltration tube to obtain a fine clean suspension without any debris.
5. Fix in chilled acetic-methanol (1:3) mixture for 30 minutes to 2 hours.
6. Centrifuge at 1000 rpm for 5 min. Discard supernatant. Repeat the process twice. After the fourth centrifugation, discard supernatant and resuspend cells in small amount of fresh fixative.
7. Draw out the cell suspension with a Pasteur pipette. Allow 2-3 drops to fall on the upper surface of a clean chilled slide. Tilt the slide. Air dry or flame dry.
8. Stain slides in diluted Giemsa stain (20 ml 0.1 Na_2 HPO_4) for 10 to 15 min. Rinse in distilled water, dry and mount in DPX mountant.

References

Hsu, T.C. (1982) *Cytogenetic Assays of Environmental Mutagens.* Allanheld Osman Publ., N.Y.

Manna, G.K., Ghosh, T.K. (1983) *Nucleus.* 26:153.

Sharma, A.K., Sharma, A. (1980) *Chromosome Techniques - Theory and Practice.* 3rd Edition. Butterworths, London.

Soares, E.R. (1972) *Mutat. Res.* 26:385.

Man as a Test System

For human systems, the indication of toxicity in terms of genetic damage is best studied from leucocytes. This system has been utilized extensively for the study of genetic polymorphism (Brusick, 1987; Preston *et al.*, 1987).

Human leucocyte test (*vide* Sharma and Sharma, 1980; Moorhead *et al.*, 1960)

Protocol

1. Draw 2-10 ml blood aseptically from the vein of exposed individuals by sterile disposable syringe and collect in a vial containing 0.1 ml heparin (5-10,000 IU/ml) sterile.
2. Sediment leucocytes by gravity at room temperature (22-38°C) for 2 hours.
3. Add 0.5 ml of plasma to 5 ml tissue culture medium aseptically. Volume of plasma depends on the white blood cell count.
4. Incubate the bottles at 37°C for 70 hours.
5. Add colchicine to 0.4 μg/ml final volume, 0.04 μg/ml or 0.1 μg/ml of colcemid or velben may be added.
6. Harvest after 2 hours. Incubate at 37°C by pouring into centrifuge tubes, spinning at not more than 600 rpm for 10 min. Add prewarmed (37°C) hypotonic solution.
7. Again incubate for 8-10 min. and recentrifuge for 10 mins.
8. Add fixative slowly to the deposit and leave undisturbed for 10 mins.
9. Shake slowly. Keep for 15 min.
10. Recentrifuge for 10 min.
11. Add fresh fixative to pellet and aspirate to mix with a long siliconised pasteur pipette.
12. Keep for 10 min. to overnight at 4°C.
13. Recentrifuge and add enough fixative to the deposit to give a suspension.
14. Centrifuge again. Add enough fixative to the deposit to give a suspension.
15. Drop on cold, clean wet glass slides drops of suspension from a height of 6 inches. Spread, blow or flame dry.

16. Stain with Giemsa and observe.

References

Brusick, D.J. (1987) Implications of treatment condition induced genotoxicity for chemical screening and data interpretation. *Mut. Res.* 189:1-6.

Moorhead, P.S., Nowell, P.C., Mellman, W.J., Battips, D.M., Hungerford, D.A. (1960) Chromosome preparation of leucocytes cultured from human peripheral blood. *Exp. Cell Res.* 20:613.

Preston, R.J., San Sebastian, J.R., McFee, A.F. (1987) The *in vitro* lymphocyte assay for assessing the clastogenecity of chemical agents. *Mut. Res.* 189:175-183.

Sharma, A.K., Sharma, A. (1980) *Chromosome Techniques - Theory and Practice.* 3rd Edition. Butterworths, London.

PART III

Sister Chromatid Exchange in Animals and Plants as Indicator of Toxicity

Sister chromatid exchanges (SCE) are cytogenetic processes related to DNA damage and repair giving an effective cytogenetic parameter for monitoring the damage (Latt *et al.*, 1982). These are due to unrepaired lesions present at the S-phase followed by chromosome replication (Wolff, 1974). It can be induced by different compounds capable of alkylating DNA. The SCE requires at least two replications for visualization in the presence of Bromodeoxyuridine (Brdu). This leads the DNA of one chromatid of metaphase chromosome to become bifilarly substituted by Brdu, the other remaining single substituted. Following the use of fluorochromes such as Hoechst 33258 and observation under fluorescent microscopy or Giemsa staining, the contrasting colour of exchanges in the two chromatids can be clearly noted. This test has also been applied to other systems including fish.

This technique for localization of SCE is quite simple, involving the usual peripheral blood culture as adopted for chromosomal aberrations excepting that Bromodeoxyuridine is added to the medium for SCE localization. The potency of this test can be estimated by the fact that it is caused by even 1/100th of the exposure required to cause aberrations (Solomon and Bobrow, 1975). It has been utilized to monitor individuals exposed to potentially dangerous compounds of the environment.

I. Protocol for SCE Test in the Chinese Hamster (Alink and van Houdt, 1985)

1. Trypsinize exponentially growing exposed Chinese Hamster cells (v-79).
2. Plate 1 ml of a suspension containing 105 cells/cc on object glasses placed in sterile 10 cm petri dishes.
3. Keep for 2 hours.
4. Add 9 ml of culture medium (Dulbecco's MEM) supplemented with 10% newborn calf serum (Gibco), penicillin (Gist-Brocades 200 units/cc) and streptomycin (Gist-Brocades 100 μg/cc). Keep for 24 hours.
5. Add to the plates 0.1, 0.2, 0.3 ml of dimelthylsulphoxide, containing extracts from 20, 40 and 60 m^3 of air.
6. After 2 hours exposure to the airborne particulate matter, replace the culture medium by

a medium containing Bromodeoxyuridine (Brdu 10 μM).
7. Continue Brdu exposure for 24 hours as the cell cycle is 12 hours. Treat the cells following the procedure of Parry and Wolff (1974) with slight modifications (Jongen *et al.*, 1981).
8. Observe the sister chromatid differential staining.
9. Score at least 50 metaphase plates per experiment per concentration.

II. Protocol for *in vivo* SCE in mice (*vide* Talukder, 1985)

1. Implant 5-Brdu tablet (50 mg each) to each etherized mouse from a specific environment, subcutaneously into the neck.
2. Inject colchicine (5 mg/kg) intraperitoneally 21-22 hours after subcutaneous implantation of Brdu tablet.
3. Sacrifice the animal by cervical dislocation 2 hours after injection of colchicine.
4. Prepare bone marrow chromosomes by usual procedure.
5. Stain the slides by fluorescence plus-Giemsa technique.
6. Stain 4-5 days old slides in 5 mg/ml 33258 Hoechst dissolved in NaCl/KCl solution for 10 min. Rinse and mount with M/15 Sorensen's phosphate buffer (pH 6.8).
7. Irradiate the slides with a 254 nm UV mineralogic lamp for 30 min. Incubate the slides in 2 x SSC (0.3 M NaCl, 0.03 M trisodium citrate) solution for 90 min. at 59°C immediately after irradiation.
8. Wash the slides thoroughly with distilled water and stain with 7% phosphate buffered Giemsa solution for 20 min.

III. Protocol for SCE Test in mice (Lin and Alfi, 1976):

1. Grow cells of mouse from affected environment in modified Eagles MEM supplemented with 10% fetal calf serum.
2. Incubate the cells in dark and grow in a medium containing 10-5 M Brdu for 18-20 or 30-32 hours (one or two cycles of DNA replication). Add colcemid (0.1 μg/ml) to culture.
3. Harvest after 2 hours by trypsinization.
4. Prepare usual air dry smear.
5. Stain with 1 μg/ml of 4-6 diamidino-2 phenylindole (DAPI) in phosphate buffered saline (pH 7.0) for 10 min.
6. Rinse in deionized water.
7. Mount in 0.2 sodium phosphate (pH 11.00).
8. Observe chromosome fluorescence with Zeiss microscope with HBO 200 w/4 Hg lamp dark field illumination, a BG 12 exciter filter and on 470 nm barrier filter.

Chromosome aberrations and micronuclei formation can be scored.

IV. Protocol for SCE in the human system

Sister Chromatid Exchange can be studied in human leucocyte culture. In the usual method for leucocyte culture, of exposed individuals, Brdu should be added 24 hours after initiation and reincubated for additional 48 hours. The culture should be grown in dark. Two hours of colcemid treatment (0.05 μg/ml) before harvesting for usual air dried preparation is needed (Pathak *et al.*, 1975; Berg, 1979; Buckton and Evans, 1982).

V. Protocol for SCE in plant test system (Kihlman and Kronborg, 1975)

1. Treat lateral roots of exposed individuals to aqueous solution containing 100 µM-5-Brdu, 01 µM 5-FdUrd and 5 µM (Urd) for 22 hours.
2. Transfer to aqueous solution containing 100 µM thymidine (dThd) and 5 µM Urd for 21 hours.
3. Treat in 0.05% colchicine for 3 hours.
4. Fix overnight in acetic-methanol cold (1:3) in dark at 20°C.
5. Rinse in 0.01 M citric acid - sodium citrate buffer (pH 4.7).
6. Incubate for 75 min at 27°C with 0.5% pectinase dissolved in same buffer.
7. Squash in 45% acetic acid. Coat with a mixture of 10:1 gelatin and chrome alum.
8. Remove cover slip by dry ice and bring the preparation to water through descending ethanol grades.
9. Incubate in moist chamber for 60 min. at 27°C with RNAse (1 mg RNAse in 10 ml 0.5 x SSC) 200 µl cover.
10. Rinse in 0.5 x SSC, stain 20 min. in H33258 (1 mg dissolved in 100 ml ethanol. Add 0.1 ml of this solution to 200 ml 0.5 x SSC). Rinse and mount in 0.5 x SSC.
11. Store for four days at 4°C.
12. Incubate for 60 min. at 55°C in 0.5 x SSC.
13. Rinse in 0.017 M phosphate buffer, pH 6.8.
14. Stain for 6-7 min. in 3% Giemsa (R66) solution in same buffer.
15. Rinse in phosphate buffer, then water, air dry, pass through xylene and mount in Canada Balsam.

References

Alink, G.M., Houdtvan, J.J. (1985) In: *Mutagenicity Testing in Environmental Pollution Control.* Zimmerman, F.K., Taylor-Mayer, R.E. (eds). Ellis Horwood Ltd, Chichester, U.K.

Berg, K. (1979) *Genetic Damage in Man caused by Environmental Agents.* Academic Press, N.Y.

Buckton, K.E., Evans, H.J. (1982) In: *Cytogenetic Assays of Environment.* Hsu, T.C. (ed.). Allanheld Osman Publ. N.Y.

Jongen, W.M.F., Lohman, P.H.M., Kottenagen, M.J., Alink, G.M., Berenda, F., Koeman, J.H. (1981) *Mutat. Res.* 81:203.

Kihlman, B.A., Kronborg, D. (1975) *Chromosoma.* 51:1.

Latt, S.A., Schreck, R.R., Sahar, E., Parks, I.J., Kittral, C. (1982) In: *Environmental Mutagens and Carcinogens.* University of Tokyo Press, Tokyo, Alan R. Liss. Inc., N.Y. 331.

Lin, M.S., Alfi, O.S. (1976) *Chromosoma.* 57:219.

Parry, P., Wolff, S. (1974) *Nature.* 251:156.

Pathak, S., Stock, A.D., Lusby, A. (1975) *Experientia.* 31:916.

Solomon, E., Bobrow, M. (1975) *Mutat. Res.* 30:273.

Talukder, G. (1985) In: *Laboratory Manual.* Centre of Advanced Study in Cell and Chromosome Research, Department of Botany, University of Calcutta. 73.

Wolff, S. (1974) Cytogenetics and environmental mutagens. In: *Genetics: New Frontiers.* Chopra, V.L., Joshi, B.C., Sharma, R.P., Bansal, H. C. (eds). Oxford and IBH Publ., New Delhi. 3, 83.

Acknowledgements

The author is thankful to Dr Sumona Mukherjee and Mr Ajoy Roy for their kind help in the preparation of the manuscript. The financial assistance from the Indian National Science Academy is gratefully acknowledged.

Botany

Leaf Yeast Survey for Air Pollution Monitoring

by P. Dowding
Department of Botany, Trinity College, Dublin 2, Republic of Ireland

Purpose of Procedure

A leaf yeast survey permits air pollution (SO_2) level mapping to a resolution of at least one square kilometre, time integrated over approximately one week. The procedure is suitable for urban or rural situations with an appropriate tree density. The method entails minimum technology and has great educational value.

Principle

Leaf yeasts are microscopic pink fungi, widespread on the surfaces of deciduous tree leaves and on leaves of grass, shrubs, and herbs. They do not harm the leaf, but exist on exudates of sugars and other nutrients. Living in this exposed position, and having efficient uptake mechanisms, these leaf yeasts are especially susceptible to inhibition by toxic substances derived from air pollution. The use of leaf yeasts has the advantage over the use of lichens in that it provides a current assessment of air pollution, and the results can be obtained within one week.

Description of method

These instructions are suitable for the use of school children in a co-ordinated survey.

1. For a survey of either urban or rural areas a suitable base map must be obtained. A suitable grid, usually 1 km, is required. This may be drawn if necessary on locally produced street plans. Enlargements of each one kilometre square can be provided for field use.
2. Visit your study square and look for the type of tree designated for your project. In the Irish situation this is ash (*Fraxinus excelsior*). Parks, gardens and roadsides will provide examples. If you wish to use a tree in a private garden, ask for the owner's consent first. Other suitable tree spp. are sycamore (*Acer pseudoplatanus*) and lime (*Tilia* spp.).
3. Locate 5 ash trees in your study square. If possible, find one near the middle of your square and others towards, but not at the edge of your square. Record whether each tree is in a field, a busy street, a quiet road, or a secluded garden. Select trees which have some leaves that are easily reached from the ground. Mark the location of the six selected ash trees on your study square map with green spots labelled A and B, C, D, E.
4. Sample the trees in the morning of the designated day. Ideally that would be Monday or Thursday (see parts 14-16). The numbers of leaf yeasts may change as autumn advances. The samples must be processed the day the leaves are collected.
5. Collect enough leaves for 9 discs from tree A (i.e., two leaves) taking one from one side and one from the opposite side of the tree and place them in a bag labelled A. The leaves should be taken from the base of long twigs. This will ensure that you have collected leaves that have been on the tree since early summer. Do not collect new young leaves from the tips of twigs. Do the same for trees B, C, D, E.
6. Bring the samples to the processing centre in five different labelled bags.

7. Prepare a clean dust-free surface by washing the bench or table with detergent and then wiping with a paper towel dipped in 70% alcohol/sterilant.
8. Cut 4 or 5 discs out of each of the leaves from Tree A using the 1 cm cutter provided. Cut the discs out of the midrib, one out of each of 4/5 leaflets onto a piece of clean paper/card. Repeat for the second leaf (see Fig. 1) making 9 discs in all. Take these 9 discs through the rest of the procedure before starting on leaves from Tree B so that you do not get them mixed up. Arrange your 9 discs with the midrib (lower surface) uppermost.
9. A pile of five petri dishes (containing 1.5% malt extract agar) is taped together with masking tape for each grid square. Carefully remove the tape as you will need this for Step 12. Take one of your petri dishes, do not open it, and label both base and top A. Then, starting with tree A discs take the petri dish and place it upside down on the bench i.e. with the agar medium in the top half. Lift the medium containing base and place it down next to the lid. Do not turn it over to look at the medium, or touch the medium, as you will contaminate it.
10. Place the petri dish lid on the leaf template provided. Deposit 9 small blobs of Vaseline on the inside of the lid in positions marked on the template. If you have a pair of forceps you will find the next step easier to perform, but they are not absolutely essential. Stick a leaf disc by its upper smooth surface to each of the nine blobs of Vaseline by pressing the disc gently onto the Vaseline, but make sure you have wiped your fingers on a clean paper towel first, or you will get the disc greasy and the experiment will not work. Lift the medium-containing base of the petri dish and replace it in the lid. Move this completed dish to one side.
11. Repeat steps 9 to 10 for the leaves collected from your other locations and you should have 5 petri dishes labelled A, B, C, D, E.
12. Make a pile of the 5 petri dishes you have filled with leaf discs (do not turn any over yet). Replace the masking tape, or use a strip of sellotape to keep the dishes together and prevent them spilling at the next stage.
13. Invert the pile of petri dishes so that the leaves in each dish are uppermost and over the agar medium. In this position leaf yeasts falling off the leaf will fall into the agar medium and grow. Place the pile of petri dishes on a table or lab bench, but not in a place where sunlight will overheat them.
14. After 24 hours invert the pile of petri dishes so that no more yeasts can fall off the leaves onto the medium. Those yeasts that have fallen will grow. Make sure you do this after 24 hours. A few minutes either side of this time will make no difference, but one hour's error could make your results out of line with the other people doing the survey.
15. Leave the stack of petri dishes at room temperature for a further 48 to 72 hours and examine on day 4 and 5. If you cannot examine them on day 4 or 5 because either would be a Saturday or Sunday, put the dishes in the bottom of a refrigerator on day 2 and take them out one or two days before you plan to count them. The leaf yeasts will go on growing in the refrigerator, but at a much slower rate than at room temperature. The pink glistening leaf yeast colonies should be visible but separate from each other on the agar under each leaf disc. These are the important yeasts. There may also be white glistening colonies and fluffy white or grey colonies, which are contaminating bacteria/yeasts and fungi respectively. Count the number of pink colonies under each disc and record the results for each leaf disc separately on your data recording sheet.

If the colonies have mostly coalesced because the dishes have been incubated for too long, reject the disc(s) on which this has happened, and put u/c (unclassified) in the square(s) on the data recording sheet. You may find it helpful if you mark the position of each colony as you count it with a felt-tip pen. For counts above 50, estimate the numbers to the nearest ten.

16. Fill in the data recording sheet and return it to the coordinator together with the study square map (Fig. 1).
17. Median values for each square should be entered on the grid map. A shading system can be devised to clarify patterns e.g.: 30+ (very good quality); 14-29 (good quality); 6-13 (moderate); 2-5 (poor); 0-1 (very poor).
18. Data from each km square can be analysed parametrically after transformation into [natural logarithm (count + 1)].

Calibration

A regression line is available between the natural logarithm of counts and mean SO_2 concentration for the previous 4 days for locations in Hamburg. This indicates that the yeasts respond to the range 0-100 ng SO_2 per cubic metre. This is suitable for monitoring the attainment of E.C. guide values for SO_2 (Dowding and Peacock, 1991).

Applications

This method has been widely applied in the E.C. alongside conventional lichen mapping and rainwater acidity measurements. There is probably potential for development of the approach to situations beyond N.W. Europe if preliminary studies are carried out on common host species and their yeasts.

References

This publication reviews the field, provides calibration data and describes applications.

Dowding, P., Peacock, J. (1991) The use of leaf surface inhabiting yeast as monitors of air pollution by sulphur dioxide. In: *Bioindicators and Environmental Management.* Jeffrey, D.W., Madden, B. (eds) Academic Press, London.

Dowding, P., Richardson, D.H.S. (1990) Leafyeasts as indicators of air quality in Europe. *Environmental Pollution.* 66:223-35.

Fig. 1. European Community Schools Air Monitoring Project

LEAF YEAST RECORDING SHEET

USE ONE SHEET FOR EACH KILOMETER SQUARE. Filled in by:

---

1. Sample site data

Country ____________ City/area__________________ Square Ref. ___________

Tree _______________ Date __/__/9 __. Time ___.___h. Disc diameter _______mm.

---

2. Notes of the environment of each tree (near main/side road, in garden, etc.)

Tree A ____________________________ Tree D ____________________________

Tree B ____________________________ Tree E ____________________________

Tree C ____________________________

---

3. Timing of operations	Time	Date
Discs cut and stuck to petriplate lids	at ____.____h	on ___/___/___
Sporefall period started (lids on top)	at ____.____h	on ___/___/___
Sporefall stopped (lids on bottom)	at ____.____h	on ___/___/___
**Stored in refrigerator from	____.____h	on ___/___/___
**Taken out of refrigerator	at ____.____h	on ___/___/___
Colonies counted	at ____.____h	on ___/___/___

**only necessary if counting not possible on Day 6.

---

4. Count data .....record your counts directly onto this sheet

Tree A									
Tree B									
Tree C									
Tree D									
Tree E									

---

5. Statistical summary

Enter the counts for each tree in ascending order, 1 to 9. If no data are missing, the count in column 5 is the MEDIAN for each tree, and the counts in columns 2 and 8 are the lower and upper confidence limits of the median. If data are missing, follow instruction N°18 carefully.

Tree	1	2	3	4	5	6	7	8	9	Lower	Upper	Median
A												
B												
C												
D												
E												

If the medians in your square are not significantly different, or if you are going to map on a km^2 basis, find the median for the whole square using the table above.

: Square median = :

Principles and Methods of Soil Algae Use for Bioindication of Soil Pollution

by E.A. Shtina[1] and M.F. Dorochova[2]

(1) Kirov Agricultural Institute, Kirov, Russia; (2) Moscow State University, Moscow, Russia

Principle

The soil accumulates different xenobiotics which are brought into it through agricultural use, from dry deposition and in precipitation. The soil is therefore the mirror of the environmental state. Especially important is the accumulation in soil of toxic substances which enter the trophic chains.

As bioindicators, the algae have some advantages compared with other microorganisms:
1. They can be easily identified as species, enabling an analysis of the algoflora of different soils.
2. The algae respond very quickly to changing soil conditions.
3. The algal cultivation is cheap and simple.
4. As phototrophs the algae react like higher plants to soil conditions.

The soil algae are used for bioindication in two ways:
1. The study of natural algal communities and the search for species or groups of algae which are characteristic of soil peculiarities or processes going on in the soil.
2. The use of particular algal species as test objects in the modelling experiments when the current state of the soil is to be analysed.

The former seems to be the more versatile and predictive, as it allows the evaluation of not only the local effect of xenobiotics but also their consequences in the near future.

The main criteria which characterize algal communities are: 1. species spectrum, 2. dominant species, 3. specific species or groups of algae, 4. spectrum of life forms.

An interpretation can be based on the absence of some definite species which are stenobiont in relation to the studied factor. The same indices can be applied to both biological analysis of soil and to indicate water pollution, namely:
1. index of species diversity;
2. index of autotrophy;
3. scale of saprobicity and trophic level.

Of special interest is the analysis of so-called "microcosms" which include a more diverse set of organisms than other algae. Under the influence of xenobiotics changes occur in the interaction of microcosm components. To analyse microcosms, it is necessary to obtain a quantitative evaluation of separate species which demonstrates its dynamics.

The coenoses on "fouling glasses", which are widely used when studying the soil algae, can be used as an example of a microcosm. On these glasses other components of soil biota may be found. In cases of soil pollution with oil, it is often observed that the content of Protozoa, especially small flagellates and naked amoeba, sharply increases. This is also true of bacteria and fungi. This may be explained by the intensive development of hydrocarbon-utilizing bacteria, the food of Protozoa. With the destruction of oil by bacteria, the algae begin to develop as well.

Another example of the use of a microcosm is to monitor soil biota when "non-traditional" organic fertilizers are applied, for instance the sediment of drainage waters. Again, both algoflora and microflora are changed.

The second direction, the autoecological aspect of indication, demands an appropriate choice of test-organism. In order to indicate residual pesticides in soil a wide range of subjects was used, ranging from microscopic algae and invertebrates up to larger animals and higher plants. In the case of algae, pollution status is determined by: quantity, rate of reproduction or disappearance, chlorophyll content, the intensity of photosynthesis, the accumulation of xenobiotics by algae and the ability of algae to decompose organic biocides.

Algological analysis of soil provides us with good results when studying the anthropogenic changes in the soil characteristics, including the early biological diagnoses of unfavourable processes.

Indication of Agricultural Impact on the Soil

The soil algae are known as bioindicators of contemporary factors of intensive agriculture, such as mineral fertilizers, pesticides, soil liming, changing water regime as a result of soil amelioration. Mineral fertilizers stimulate the development of algae in many cases. But long-term application of some fertilizers can result in consequential pollution effect: for example, in the soil acidification resulting from proton exchange when ammonium ions are absorbed. This decreases the amount of algae and results in the complete disappearance of the blue-green algae.

The Effect of High Doses of Mineral Fertilizers
(up to N800 P300 K880 kg/ha of an active substance)

In the field experiment, doses of up to 430 kg/ha inhibited the development of algae only in the initial stages. Higher doses of fertilizers affected different algae groups in various ways. A positive correlation (coefficient +0.74) between the dose of nitrogen fertilizers and the quantity of algae was found for the single-cell green algae. On the contrary, higher doses of K (600 kg/ha and more) inhibited the green algae. The blue-green and diatomic algae responded positively only at the relatively lower doses of nitrogen and phosphorus. The nitrogen-fixing algae disappeared in the face of 200 kg/ha of N. On the other hand, any K-concentrations used in the experiment (up to 600 kg/ha) stimulated the development of blue-green algae, with densities of 15-38 million cells in 1 g of epigeal sand layer.

Of interest is the fact that the yield of oats in this experiment was highest when fertilizer doses did not exceed N300 P170 K110 kg/ha. These data suggest the possible selection of algae to indicate toxic doses of mineral fertilizers, taking into account the different soil types.

The traditional organic fertilizers under field conditions in very high doses (up to 1000 t/ha of manure) did not negatively affect the amount of algae, increasing the proportion of blue-green and diatomic algae in derno-podsolic soils.

It can be seen that the application of fertilizers can change the structure of algal groups in soil. Chemical improvement of soil influences the content of algae: liming increases the amount and the activity of nitrogen-fixing blue-green algae and decreases the proportion of green algae. On the contrary, when the saline soils (solonetz) are ameliorated with gypsum or sulphuric acid, the dominant blue-green algae are replaced by greens.

When sewage was utilized for irrigation, the waste waters from the treatment plant depressed the development of algae, and the degree of depression depended on the dose and the duration of irrigation (up to 1.5 years) (Table 1).

Of special interest is the accumulation of various biocides in the soil. The algae, as phototrophs, are very sensitive to herbicides as they react selectively, like higher plants. There are many data available on the influence of various pesticides on algae; sometimes the effect is not direct, as they influence the heterotrophic microflora or invertebrates.

Metabolic reactions of algae to herbicides include: the decrease in chlorophyll content; the reduction of dry weight of cells; the disturbance in the synchronization of cell division; the inhibition of nitrogen-fixing activity in blue-green algae. Some species can accumulate the pesticides and then detoxify them.

Influence of Industrial Wastes on Soil Algae

Algae may also be used as indicators of toxicity on industrial land-fill sites. They are first-colonizers, and demonstrate whether or not the site may be suitable for agricultural use.

The algological analysis gives hopeful results during the study of anthropogenic changes in the soil as well. There are data on the reaction of soil algae on some industrial wastes: on local pollutions (for example, oil floor and the flood of confined waters during oil-extraction and of mine-waters during coal extraction) or on more total pollution caused by smoke-gas emission. It is shown that even slight oil pollution of the soil decreases the species diversity of algae. High oil pollution reduced their number to the value not revealed directly by conventional methods. The yellow-green and diatomic algae, which practically disappeared even on the slightly polluted area, turned out to be more sensitive to oil pollution (Table 2).

During the process of oil decomposition some algal communities are restored, first of all green and blue-green nitrogen-fixing algae appear. The highly mineralized waters from deep strata, which are pumped to the surface and discharged during oil extraction, also influence the algae. In particular, the development of blue-green algae is stimulated. Cases are known of derno-podsolic soils which are salinized by industrial discharge, develop technogenic saline and become covered with massive films of blue-green algae (Oscillatoriales). The yellow-green algae are very sensitive to oil and salt pollution of the soil. Table 2 gives a picture indicating the significance of some algal groups.

The mineralized waters (sometimes thermal ones) vented during oil extraction, causing disturbances of mineral, water and thermal regimes (down to the formation of technogenic solonchaks) sharply change the soil algoflora. Under the influence of chloride-sodium salinization, yellow-green and many green algae disappear but the development of blue-green, non nitrogen-fixing species increases; in some areas a solid algal crust is formed on technogenic solonchaks. Such a crust is similar in composition to takyr (clay soil in desert and semi-desert regions). As on the takyrs, the blue-green algae become the pioneers of organic matter formation.

The impact of acid mine-waters was observed in the subzone of the southern taiga. Under the influence of highly acid mine-waters (pH 1-1.5) and filtrating waters from refuse and mine dumps with high concentrations of sulphuric acid, ferrum and aluminium sulphates, the higher vegetation (coniferous forests) perish, soils are overlapped with phytotoxic technogenic deposits (2-24 cm in thickness), and acid solonchaks are formed. In spite of the apparent lifelessness, the technogenic substrate is occupied by soil algae, microscopic fungi and protonema of mosses. The algal composition is very peculiar; small diatomea and unicellular green algae sharply dominate. On moist areas, the multi-layered mats of the green filamentous algae (of rare occurrence in soils) were observed (*Zygogonium ericetorum* Kütz).

The soil algae are distributed throughout the areas (which are under the influence of acid filtrating waters) and were found even in close vicinity to refuse and mine dumps. In the lower part of the technogenic flow the amount of living soil algae varies from 106.5 thousands to 5.95 millions of cells per 1 cm^2; including the number of diatomic algae - from 90 thousands to 5.6 millions of cells; the number of green ones - from 16.4 to 310.8 thousands of cells per 1 cm^2. Their biomasses are 0.3-4.6 mg/cm^2, and the biomasses of protonema - up to 4 mg/cm^2. On the green algae and protonema covers the so called "feeding spots" of microscopic fungi and small invertebrates appear.

The enumerated spore bearing plants form peculiar phytocenosis which is the first stage of succession. The protonema of mosses and soil algae are producers of organic matter and initial links of the food chains. Contrary to the earlier described cases of overgrowing of technogenic substrates, in the given case a new type of biological invasion is revealed; the protonema of mosses together with its small bryophilous diatomic algae are the pioneers.

In connection with the fact that about 1 million tons of agents of acid character are emitted into the atmosphere, they inevitably affect the different soil properties. The occurring acidification of soil and depression of higher plants are well known. For the first time the influence of air pollution on the development of soil algae was investigated. The investigations were carried out near copper-smelting plants in the middle taiga and in the steppe zone. In the forest zone wastes of metallurgical plants depressed the soil algoflora, mainly blue-green and yellow-green algae. In the steppe zone the impact of acid wastes on algae is more expressed. The algoflora of the steppe zone, represented mainly by blue-green algae, is destroyed. The degree of inhibition depends on the distance from the pollution source (PS) and on the wind-rose. Downwind, even at the distance of 15-20 km from PS, only green and diatomic algae remained, being accompanied by a mass of Protozoa. The grass vegetation in this case did not show any sign of depression. Only on one experimental plot (where north winds occur only 3% days per year) the typical steppe algoflora remained with dominance of the blue-green algae.

Thus the soil algae turned out to be more sensitive to acid discharges as compared with higher plants. The absence of blue-green algae could be considered as a bioindicator of soil acidification due to the industrial wastes. The preliminary analyses of algoflora of polluted soils allows us to distinguish some algal groups which indicate certain anthropogenic soil changes (Table 3).

Table 1. Influence of industrial waste water on soil algae

Amount of cells in thousands per 1 g soil				
	Depth 5 cm		Depth 20 cm	
Time (date)	Control	Experiment	Control	Experiment
26.09.74*	5500	5100	-	-
30.09.74	3100	79	-	-
04.10.74*	1100	330	660	630
07.10.74	7000	250	880	550
25.10.74*	7500	160	3300	19
29.10.74	5100	79	2100	460
01.11.74*	6300	330	1000	120
04.11.74	6000	80	2500	80
20.06.75	2700	72	430	52

* the time of irrigation (watering)

Table 2. Influence of oil pollution on algoflora

Number of algae species					
Degree of pollution	Blue-green	Green	Yellow-green	Diatomic	Total
Control ("clean" sand)	31	28	8	7	74
Slight pollution	14	10	-	-	24
Average pollution	6	4	1	2	13
High pollution	-	1	-	-	1

Table 3. Groups of soil algae indicating environmental pollution

Soil processes	Intensification of development	Weakening of development
Salinizsation	Cyanophyta	Xantophyta
Oil pollution		All of the groups
Acid pollution with mine-waters	Bacillariophyta	Cyanophyta, Xantophyta
Acidification of soil		All of the groups, especially Cyanophyta
Metal accumulation		All of the groups, especially Cyanophyta
Pollution with communal wastes	Cyanophyta, Bacillariophyta	Xanthophyta
Accumulation of pesticides		Xanthophyta

References

Shtina, E.A., Negapova, L.B. (1977) Changes of algoflora under the influence of soil pollution. Thesis of the Fifth Conference of the USSR Soil Science Society. Minsk. pp. 264-266 (in Russ.).

Shtina, E.A. (1979) Soil algae as bioindicators of environment pollution. *Proceedings of the Fifth Conference on Lower Plants in Transcaucasus*, Baku. "Alm" Publishing House. p. 39.

Shtina, E.A., Hollerbach, M.M. (1980) Principles and methods of using soil algae for bioindication. In: *Problems and Methods of Biological Diagnostics and Soil Indication.* Moscow State University Publishing House. pp. 75-84 (in Russ.).

Shtina, E.A. (1981) Soil algae as bioindicators of environmental state. In: *Problems of Soil Science.* Sov. pedobiologists to the XII Int. Congr. of Soil Sci., "Nauka" Publishing House, Moscow. pp. 84-88.

Shtina, E.A. (1985) Soil algae as pioneers of overgrowings of technogenic substrates and indicators of damaged land state. *J. Gen. Biol.* 46:435-443 (in Russ.).

Shtina, E.A. (1987) Microscopic algae as indicators of soil pollution with toxic substances. *Proceedings of Res. Inst. of Agricultural Microbiology.* p. 44-45b (in Russ.).

Shtina, E.A. (1988) Algae as indicators in estimation of soil pollution. In: *Ecotoxicology and nature protection.* Riga. pp. 216-218 (in Russ.).

Zoology

Use of Imposex in Marine Gastropods as an Indicator of Tributyltin (TBT) Contamination

by G.M. Hughes
Research Unit for Comparative Animal Respiration, University of Bristol, Bristol BS8 1UG, U.K.

Principle

The term 'imposex' describes the superimposition of male characteristics on to females of certain gastropod molluscs. It has been recognized in natural populations and there is now substantial evidence that its increased occurrence has arisen because of the use of antifouling coatings containing Tributyltin (TBT) compounds not only on boats but also on fish farm cages.

Recent studies suggest that imposex may be initiated by as little as 1 ng/l of Tributyltin in natural populations where it develops over long periods. But in laboratory situations specimens exposed in experimental tanks for periods as short as one month can indicate concentrations down to 20 ng/l.

Procedures

For practical use it is first necessary to recognize the two sexes of a suitable stenoglossan gastropod such as *Nassarius obsoletus* or *Nucella lapillus*. As many present day populations are affected by imposex, the penis as a male characteristic is no longer reliable. The best character to differentiate between the sexes appears to be the presence in the female of a sperm-ingesting gland called the 'brown gland' which is situated immediately posterior to the capsule gland. It often has a reddish tinge and its size and colour does not appear to vary with the reproductive state of the female.

Suitable indices have been described for quantifying the degree of imposex of which relative penis size (RPS) is the most useful. Length of the penis is easiest to measure but does not give a true impression of the difference in mass. Accordingly as weight or volume is related to the length cubed (Bryan *et al.*, 1986) the RPS values are calculated as:

$$\frac{(\text{mean length of female penis})^3}{(\text{mean length of male penis})^3} \times 100$$

An RPS value of 50% means that the mean penis size of the female has half the bulk of that of the male.

In addition a *vas deferens* sequence (VDS) index has also been described (Gibbs *et al.*, 1987).

It seems probable that suitable gastropods may be used as indicators in different countries, and perhaps some may be sensitive to pollutants other than TBT.

References

Bryan, G.W., Gibbs, P.F., Hummerstone, L.G., Burt, G.R. (1986) The decline of the gastropod *Nucella lapillus* around south-west England: evidence for the effect of tributyltin from antifouling paints. *J. Mar. Biol. Ass. U.K.* 66:611-640.

Gibbs, P.F., Bryan, G.W., Pascoe, P.L., Burt, G.R. (1987) The use of the dog-whelk, *Nucella lapillus*, as an indicator of tributyltin (TBT) contamination. *J. Mar. Biol. Ass. U.K.* 67:507-523.

Soil Animals as Biomonitors of the Quality of Woodland Environment

by Andrzej Lesniak
Institute of Zoology, The Polish Academy of Sciences, ul. Wilcza 64, Warsaw, Poland

Principle

To determine the extent of degradation of forest communities affected by human influence, including pollution effects but not excluding other influences.

It seems essential to draw a distinct line between bioindication and monitoring; the purpose of the latter cannot be limited to estimations of pollution loads on the basis of biotests, but it must comprise evaluation of the state of a certain part or element of the biosphere. Obviously, the state of the biosphere does not depend only on the degree of its pollution but on other factors. Accordingly, monitoring should be concerned with the most fundamental elements of the biosphere which are subject to human management.

Health conditions and productivity of forests depends on the normal state and functioning of all their cardinal components. One of these elements is soil fauna, whose reduction and elimination (Gorny, 1975) considerably sets back the rate of tree stand production. For this reason works on monitoring were planned out and initiated, focusing on evaluation of the state and dynamics of changes in soil animals in forests of Poland. The studies were carried out in stands of pine in the habitats of young pine woods (Leucobrio-Pinetum and Pseucedano-Pinetum) of the third age class and the third habitat quality class. This choice was justified by the following reasons:

1. Monitoring must be performed throughout many successive years and only forests provide adequate conditions for conducting such a long-term observation, undisturbed by any changes in the economic management regime or in plant species composition (crop rotation).
2. Due to many compensating mechanisms, the forest is one of the most natural elements of the environment and is subject to smaller quantitative and qualitative oscillations in the soil fauna composition.
3. The influence of meteorological factors is lower in forests than in agriculture. Additionally those factors obscure the causes of changes in dynamic processes advancing in soil animal populations.
4. The choice of the tree stand category was based on the following premises: nowadays in Poland young pine woods with pine trees are the most common type of habitat and the majority have tree stands ranking among the third age class and the third habitat quality class.

Method

Studies on soil animal monitoring were located in 50 forest inspectorates on regular standard soil plots. Out of these 50 plots, ten were located within the first (lowest) zone of pine needle sulphur content (according to the map worked out by the Botanical Garden of the Polish Academy of Sciences (Molski, 1985)), fifteen were located in the second, fifteen in the third, and ten in the fourth zone (Silesia, the vicinities of Opole). The Carabidae and Scarabaeidae communities have been studied. The material has been sampled in Barber traps modified by Szyszko (1985). The selected taxonomic groups were chosen for the following reasons:

1. These easily trapped groups are particularly abundant in samples, they are common, diversified and their systematics as well as ecology comparatively well known.
2. They perform a significant and diversified function in forest biocenoses, as Carabidae communities comprise:
 - large zoophages controlling abundance of the so-termed forest primary pests, i.e. phytophages;
 - small zoophages controlling abundance of numerous organisms in the detritus trophic chain,
 - partly phytophagous hemizoophages.

On the other hand, the most numerous trapped soil species of Scarabaeidae perform a significant function in forests due to a specific organic fertilization (burying litter, amounting to some several hundred kilograms per hectare). Furthermore, what is very important, these animals are a very sensitive indicator of soil pollution.

The analysis will include other groups of soil surface animals sampled in the employed traps (Diplopoda, Isopoda, Pseudoscorpiones, Aranei, Opiliones). Depending on the abundance of the sampled material, the prospective analysis will be more or less thorough.

a) Field studies

The criteria which form the choice of the regular research plots have already been stated. 300 Barber traps modified by Szyszko (1985) assigned for temporary sampling were deployed on the plots.

b) Work out - materials and results

Animals were cleaned and divided into appropriate taxonomic groups, namely, Carabidae, Scarabaeidae, Isopoda, Diplopoda, Pseudoscorpiones, Aranei, Opiliones, Coleoptera varia and others. The Carabidae were determined to species level. Their abundance and the abundance of the species *Geotrupes stercorosus* (Scar.) were determined.

Results

The procedure described above allows parameters to be determined which indicate the degree and direction of decline in function of animals significant in ecological productivity. The following table lists these parameters and states the effect of intensifying pressure.

Ecological parameter	Effect of increased pressure
Species diversity	Decrease in diversity
Abundance and biomass of communities examined	Diminution of abundance and total biomass
Dominance structure of communities	Increase in the contribution of dominants to the whole community
Structure of community functions	Increase in the contribution of small zoophages to the community of Carabidae and, under extreme intensification of pollution effects, also an increase in the contribution of hemizoophages
Structure of community	Decrease in the H value entropy estimated after the Shannon-Wiener formula
Phenotypical diversity of the *Geotrupes stercorosus* population	A greater diversity of population

The above trends were noted in numerous studies carried out in the Institute of Ecology, PAS (Prot, 1985) as well as by the author (Lesniak, 1971, 1979, 1980, 1986). The investigations which were initiated in Poland in 1986 have been used since for forest management. This confirmed the usefulness of continuing the observations.

References

Gorny, M. (1975) Zooekologia gleb lesnych. PWRiL Warszawa p. 311

Lesniak, A. (1971) Zmiany skladu i struktury zgrupowan Carabidae Col. w poblizu Zakladow Azotowych w Pulawach. Dok. IBL Warszawa p.17.

Lesniak, A. (1979) Mozliwosci bioindykacji antropogenicznych zniekszatalcen srodowisk lesnych na podstawie zmian w zgrupowaniach bezkregowcow. Materialy I Sympozjum "Ochrona ekosystemow lesnych". Wyd. SGGW-AR Warszawa pp. 5-13.

Lesniak, A. (1980) Badania zmian w strukturze zgrupowan entomofauny naziemnej jako wzkaznikow zniekszta1cen edosystemow lesnych. Dok. IBL, p. 144.

Lesniak, A. (1986) Methods applied in global monitoring of forest ecosystems. In: *Proc. European MAB Conference on Biosphere Reserves and Ecological Monitorings.* Ceske Budejovice, Czechoslovakia pp. 171-175.

Molski, B. (1985) Informacja na temat zanieczyszczenia atmosferya w Polsce i ich skutkow gospodarczych. Dok. Ogrod Botaniczny PAN, Warszawa.

Prot, E. (1985) Monitoring ekologiczny ekosystemow ladowych mozliwosci i zakres. *Wiad. Ekol. T.* 31, z 1, pp. 55-64.

Szyszko, J. (1985) Pulapka STN do odlowow Carabidae. Prace Kom. Nauk. PTG nr 91. Warszawa pp. 34-41.

Zoocenotic Characteristics in Monitoring the Effect of Pollution in Terrestrial Ecosystems

by L.I. Khotko[1] and T. I. Artemieva[2]

(1) Institute of Zoology, Byelorussian Academy of Sciences, Minsk, Byelorussia

(2) Institute of Biology, Kazan Branch of the Russian Academy of Sciences, Kazan, Russia

Principle

The studies of zoocenotic characteristics such as population density, trophic structure and species composition of soil invertebrates inhabiting the major plant formations are aimed at obtaining the definite relations and parameters characterizing the state of soil invertebrate communities most sensitive to industrial emissions of oil chemical enterprises and oil production waste water.

The studies are based on comparing certain parameters characterizing the vital activities of soil invertebrates in biogeocenoses under pollution with control areas situated far from the pollution sources. In parallel, agrochemical assay of the soil samples, which are to be collected from every biogeocenosis layer-by-layer in four replications to a 30 cm depth, should be done.

Sampling is to be performed three times a year (in spring, summer and autumn) to eliminate any seasonal variations.

In each biogeocenosis 8 to 10 samples should be taken preliminarily for mesofauna and microarthropods. To obtain significant population densities and to determine the least sample size the formula $N = (tV/P)^2$ is used where N = least sample size; V = coefficient of variation (%) defined by the expression $V = \sigma/M \times 100$; P = required experimental accuracy (%); t = Student test; σ = mean square deviation; M = arithmetic mean.

Soil samples for microarthropods are taken by a steel drill (with a 25 cm^2 cross-sectional area). It is recommended to take soil samples for mesofauna by digging layer-by-layer, which is followed by handsorting; Barber's traps may be used too. At least 20 samples for microarthropods should be collected at a depth of 5 cm where 90 to 95% of small arthropods are concentrated. They are to be extracted from soil samples using light or heat extractors of any design. While extracting the soil microfauna, the following requirements should be met: the sample is placed into the extractor top down; the surface temperature of the sample should not exceed 27°C at the beginning and 45°C at the end of extraction. The extraction usually lasts 5 to 7 days, but for samples of a high moisture content the time may be increased until they become fully dry. The material is fixed in 70% ethyl alcohol mixed with glycerine (8:2).

In counts of mesofauna the sample sizes should be most suitable for the zone studied (50 x 50 cm to a depth of 40 cm). Glass jars (0.5 litre) or polystyrene vials should be buried in the soil in batches of 25, equally spaced from one another. The traps remain buried during the whole growing season. The procedure used in the study should include joint investigation of various pedobiont groups both in space and in time, which necessitates dynamic observations over the same plots periodically for several years.

The species diversity index (H', in $n_i$ts) is to be found from the formula suggested by Shannon and Weaver (1949) :

$$H' = \frac{1}{N} (N \ln N - \sum n_i \ln n_i)$$

where N = the total number of individuals of all species; n_i = the number of individuals of the i-th species. The standard error of the Shannon index ($m_{H'}$) and the number of degrees of freedom are calculated from Hutcheson's (1970) formula:

$$m^2_{H'} = \frac{1}{N}\left[\sum \frac{n_i}{N} \ln^2 \frac{n_i}{N} - (H')^2 \right] + \frac{(S-1)}{2N^2} + ...$$

where S = number of species;

$$v = N_1 N_2 (m^2_{H'_1} + m^2_{H'_2})^2 / (N_2 m^4_{H'_1} + N_1 m^4_{H'_2})$$

where v = number of degrees of freedom.

For determining the significance of the differences in the diversities of biotopes compared the Student t-test may be recommended:

$$t = (H'_1 - H'_2) / \sqrt{m^2_{H'_1} + m^2_{H'_2}}$$

The proportional distribution of the number of individuals among species may be estimated following Gray (1979). If the abscissa is geometrical classes of the number of individuals per one species (1, 2 - 3, 4 -7, 8 - 15, etc. individuals), and the ordinate is the cumulative percentage of their abundance, the resultant line will be straight or broken. For stable communities the resultant will be a steep straight line (a small number of classes of the number of individuals per one species). The broken line characterizes unstable communities, in which the number of classes of the number of individuals per one species is larger. The method allows the disturbed communities to be revealed, and the species, which are indicators of the altered conditions, to be selected.

The dominance structure of the invertebrates may be studied following Stöcker and Bergmann (1977) and Stöcker (1977). Particular cumulative Shannon's indices (h_i) reflect the dominance structure of soil invertebrates in various biogeocenoses. If the abscissa is the number of species and the ordinate is the cumulative species diversity indices for a given number of species, the resultant curve will be exponential with the limit value H'. This curve may be approximated by the e-function of the type $y = a+be^{cx}$, where $y = h$, $x = S$ (number of species) and $a \approx |b| \approx H'$. Using the parameter c, which is negative in all cases, one can predict the stucture of community from the proportion of dominants within one habitat and from their changes.

Evenness, which is meant as the degree of uniformity of the abundances of different species comprising the community, is defined by the ratio of the species diversity index to its maximum value. The latter is a natural logarithm of the number of invertebrate species.

The model of the dominance structure composed simultaneously gives an idea of three interrelated parameters of the community, namely, the number of species, the species diversity indices and evenness. The species diversity index is closely related with the number of species. In disturbed communities it is inversely proportional to the degree of dominance. Evenness of the community is directly proportional to bending of the dominance structure curve with the same value of the species diversity indices and inversely proportional to the degree of dominance.

The long-term studies carried out to investigate the effects of soil pollution with oil and oil-production waste water on the invertebrate complex have revealed that (1) oil and oil-containing waste water have a long-lasting negative effect on the soil animals (collembolans, Gamasina, Acaridiae, Tarsonemini, Oribatei), almost fully eliminating them within the obligate zone of pollution and drastically decreasing their number over slightly polluted areas. The decrease of the pedobiont abundance is accompanied by depletion of their species composition; (2) the natural restoration of the soil animal complex is a slow process, which distinctly correlates with the rate of oil decomposition in soil, the decrease of toxic salt contents in it, the activity of microflora and direction of technogenic plant succession; (3) a higher rate of forming the small invertebrate complex as compared to a large invertebrate group is a general specific feature; (4) small soil animals (microarthropods) are indicators of relatively rapid changes, whereas large ones (mesofauna) suggest deep transformation of oil hydrocarbons.

Soil pollution with industrial emissions leads to the substitution of dominants, disturbs the species proportion in their abundances, lowers the ecological pyramid due to all trophic groups. In the invertebrate communities the regulatory role becomes shared by a smaller number of species (by 1.5 times on average). Ecologically stenovalent species are progressively replaced by euryvalent ones. Carabidae, Staphylinidae, Elateridae and Curculionidae communities are most demonstrative in this respect.

As a controllable characteristic for estimating the conditions of communities a dominance structure was chosen, which requires a dynamic study of the following parameters:

1. species diversity indices;
2. the degree of dominance;
3. percentage of rare species;
4. the proportional distribution of the number of individuals among the species.

It should be noted that the characteristics just enumerated tend to decrease under the action of industrial emissions and oil-production wastes. Even slightest deviations from the background conditions point to some disturbances. So, it is shown that in the pine forests, which were affected by emissions of the oil-chemical enterprises and are more sensitive to any pollution, a decrease of the total number of soil invertebrate species frequently results in diminishing the species diversity indices and one of its components, namely, evenness. In the oak forests emissions of the oil-chemical enterprises slightly reduce the number of Staphylinidae species (by 1.3 times) and somewhat increase the number of rare species. The species diversity indices for the Staphylinidae communities (1,1406 - 1,2522) and the structure of the proportional distribution of individuals among species remain, however, unchanged. It has been noted that in some biotopes containing equal numbers of the species Elateridae, the species diversity indices are reduced (from 1,4063 ± 0,06467 to 1,1348 ± 0,11936), a fact which may be ascribed to non-uniformity of the proportional distribution of individuals among species. Hence, the index of evenness is small (0.58).

References

Gray, J.S. (1979) The development of a monitoring program for Norway's coastal marine fauna. *AMBIO*. 8(4):176-179.

Hutcheson, K. (1970) A test for comparing diversities based on the Shannon formula. *J. theor. Biol.* 29(1):151-154.

Shannon, C.E., Weaver, W. (1949) *The Mathematical Theory of Communication*. Urbana: Univ. Illinois Press. 117 pp.

Stöcker, G. (1977) Ein Modell der Dominanzstruktur und seine Anwendung. 2. Bioindikation, allgemeine Ergebnisse. *Archiv für Naturschutz und Landschaftsforschung*. 17(2):89-118.

Stöcker, G., Bergmann, A. (1977) Ein Modell der Dominanzstruktur und seine Anwendung. 1. Modellbildung, Modellrealisierung, Dominanzklassen. *Archiv für Naturschutz und Landschaftsforschung*. 17(1):1-26.

Laboratory Culture of *Gammarus pulex* as a Test Organism for the Assessment of Pollutant Toxicity

by C.P. McCahon and D. Pascoe
School of Pure and Applied Biology, University of Wales, College of Cardiff,
P.O. Box 915, Cardiff CF1 3TL, U.K.

Purpose of Procedure

Specimens for use in aquatic toxicity testing come from two main sources, field collections and laboratory cultures. By using a laboratory culture the investigator is assured that the test animals are reared in water of controlled quality, and have not been previously exposed to a toxicant, a situation which can increase or decrease an animal's resistance to a second challenge of toxicant. Nor are such animals subjected to the handling stresses, involved in collection and sorting before a test, which may increase sensitivity towards a pollutant.

Although the routine and standardized laboratory culture of several fish species for use in toxicity tests is widespread, the only invertebrate for which this is the case is *Daphnia* spp. The main reason for this appears to be the lack of successful culture methods for many invertebrate species and the difficulty of even maintaining field-collected specimens in the laboratory for any appreciable time. However, it is important that standardized culture methods be developed for a range of invertebrates in order that the battery of tests for assessing pollutant toxicity on the aquatic ecosystem may be expanded.

Members of the Amphipoda genus *Gammarus* are widespread in Europe, Scandinavia and N. America and the following is a description of a simple method for the laboratory culture of *Gammarus pulex* (L.), an important freshwater detritivore and fish food organism which is known to be among the most sensitive of several fish and invertebrate species to a range of pollutants (McCahon and Pascoe, 1988a,b).

Culture Technique

Specimens of *G. pulex* should be collected, using hand nets, from an unpolluted source and transported to the laboratory in aerated and, if necessary, ice-cooled water. Diseased and parasitized specimens should be rejected.

Since gravid females may become attractive to males and enter pre-copula at or shortly before the hatching of the eggs, both visibly gravid females and pairs in pre-copula are removed from the collected sample. These are placed in a 1 litre plastic container (approximately 170 x 170 mm) with the base replaced by nylon mesh (0.5 mm aperture) through which juveniles can pass after release from the brood pouch. This "breeding container" is then suspended in a rearing tank (2-8 l) with a flow-through supply of dechlorinated, aerated tap water. Adults can be fed on a variety of leaves from species, such as horse chestnut (*Aesculus hippocastanum* L.), elm (*Ulmus* spp.), oak (*Quercus* spp.) or sycamore (*Acer pseudoplatanus* L.). Leaves should be collected during the autumn fall, air dried and stored until shortly before required, when they are "conditioned" for at least 10 days in organically enriched dechlorinated water to initiate microbial breakdown since bacteria and fungi are important components of the gammarid diet.

Provided sufficient gravid females (> 200) are originally placed in the "breeding container", after 5 days it can be removed and a large number of newly hatched individuals will have been released from the brood pouches and fallen through the mesh of the breeding container into the rearing tank. These juveniles are supplied with conditioned leaves for cover and also to graze upon. Juveniles also feed upon adult faeces and after removal of the breeding container these can be supplied until the animal can feed entirely upon conditioned leaves, i.e. after about 25 days.

By transferring the container of breeding adults to fresh rearing tanks at periodic intervals, groups of animals of different age ranges as well as newly hatched juveniles are available for toxicity tests.

Culture Characteristics

When juveniles are released from the brood pouch they possess 5 segments on the primary flagellum of each antenna, and this number increases as growth progresses. When the number of antennal segments or mean body length is plotted with animal age (Table 1) a typical sigmoid growth curve is obtained and from this the age of cultured specimens can be calculated.

Table 1. Mean body lengths and mean antennal segment counts (± standard deviations) for 40 *G. pulex* of several age groups.

Age (Days)	Mean body length (mm)	Mean number of antennal segments
1	2.04 (±0.09)	5.00 (±0.00)
7	2.17 (±0.19)	5.48 (±0.52)
25	2.90 (±0.31)	6.71 (±0.97)
47	4.03 (±0.51)	9.76 (±1.18)
67	3.95 (±0.34)	9.86 (±0.93)
82	8.16 (±0.85)	19.94 (±2.14)
110	8.59 (±1.27)	19.37 (±2.13)
170	8.70 (±0.67)	20.20 (±1.92)
220	8.16 (±0.85)	19.94 (±2.14)
469	11.46 (±1.70)	26.81 (±4.64)
unknown (field collected)	9.48 (±0.90)	22.00 (±2.65)

Although an increase in the culture temperature will increase growth rate, a count of antennal segments can still be used to estimate age to within several days if the original "calibration" curve was prepared using data obtained at the appropriate culture temperature.

Approximately 70% of cultured juveniles survive to reach sexual maturity within 130 days at 13°C. Sexual maturity is reached at 14-16 antennal segments, after moult 10 when in the male the genital papillae are visible and in females the oostegites are fully developed with long fringed bristles which interlace with one another to form the brood pouch.

The ability of females to produce 2-5 broods containing a mean of 16 eggs (range 10-26) ensures that a large number of offspring can be reared from one group of adults and, by increasing temperature and providing excess food under laboratory conditions, it is possible to culture animals throughout the year and to reduce the time taken to reach sexual maturity.

Conclusions

G. pulex is readily amenable to laboratory culture, using a simple, inexpensive technique which requires minimal space and maintenance. Large number of offspring are produced which are available throughout the year at any required developmental stage.

In view of its sensitivity to a range of pollutants, *G. pulex* can be used in the laboratory to perform toxicity tests on novel chemicals or effluents and also for the *in situ* assessment of effluent toxicity in the field.

References

McCahon, C.P., Pascoe, D. (1988a) Culture techniques for three freshwater macroinvertebrate species and their use in toxicity tests. *Chemosphere* 17:2471-2480.

McCahon, C.P., Pascoe, D. (1988b) Use of *Gammarus pulex* (L.) in safety evaluation tests - culture and selection of a sensitive life stage. *Ecotox. Env. Safety* 15: 245-252.

In situ Assessment of Effluent Toxicity Using Fish and Invertebrates

by D. Pascoe and C.P. McCahon
School of Pure and Applied Biology, University of Wales, College of Cardiff,
P.O. Box 915, Cardiff CF1 3TL, U.K.

Purpose of Procedure

Many effluents discharged into river systems comprise complex mixtures of chemical waste, for which, in most cases, reliable methods of chemical analysis are not available and in others even the constituents which make up the waste may be unknown. In such situations the detection and even quantification of toxic effects may be carried out by recording the responses of animals exposed within the water body, into which the effluent is discharged.

Methodology

(see Pascoe and Edwards, 1989)

Selection of test species

Species used in the test should be relevant to the particular river system under investigation and if resources permit, should include representatives of annelids, crustaceans, insects, molluscs and fish. This will ensure not only examination of a spectrum of taxonomic groups but also selection of species normally found (I) in the water column, (II) on the river bed and (III) in the substratum.

The chosen species may be obtained:

1. By normal sampling procedures (e.g. kick-sampling, netting) from (a) uncontaminated sites upstream of the effluent discharge; (b) from another non-polluted river of similar water quality.
2. From stock cultures maintained in the laboratory. These animals will require acclimation to the appropriate water quality conditions (pH, temperature, hardness) prior to testing.

Design of test containers (see Ormerod *et al.*, 1987)

Fish cages may be constructed using a metal or rigid plastic framework and netting of an appropriate size for the species used. Alternatively plastic "drain" pipes, with one fixed and one detachable (for observation and removal of dead animals) meshed end piece may be used. Cages or pipes should be staked in position and if the depth of water permits, a mid-water position will significantly reduce problems caused by silting up of the containers.

For invertebrates, small plastic containers provided with mesh "windows" allowing free circulation of water are normally adequate. Inclusion of sediment will increase the realism of the test in the case of benthic species but it may make observation of individual animals difficult. More sophisticated systems involving artificial substrates colonized with known species may also be employed to detect the effects of contaminants.

The precise design of containers will vary to some extent with the species used but the following guidelines should be noted:

1. Containers should be sturdy enough to withstand adverse conditions such as those occurring during a spate. Several small invertebrate containers may be placed within a single larger vessel securely staked or weighted to the river bed.
2. Water must pass freely through the test container in order that animals are exposed under relatively natural conditions. Excessive clogging of mesh or netting due to leaves, debris or silt must be avoided.
3. The design should facilitate ease of observation of test animals and the removal of dead ones. While this may most easily be accomplished by use of open-top vessels, it must be noted that such a system is also susceptible to flooding, and a flotation device must be incorporated which will raise the container during a spate.

Positioning of containers during a test

Effects of the effluent will clearly vary with the degree of dilution and consequently with the position of the test containers. Ideally control animals should be placed in an uncontaminated stretch of water above the discharge level and test animals at, and at varying distances below, the discharge level. A graded series of concentration-related responses may then be detected.

Recording toxic effects

The investigator must have a clear idea of the toxic effects which will be recorded for each species and this will be based upon laboratory tests using a range of pollutant types with each of the chosen test species.

Lethal toxicity

The most dramatic response and the most widely employed in toxicity investigations is mortality and although for most species this can easily be detected, for some, e.g. molluscs, considerable experience is required.

Sub-lethal toxicity

Many effluents may be toxic but not acutely lethal and in such cases sub-lethal indications of toxicity are required. There may, for example, be changes in (a) feeding rate (based upon the weight of supplied food consumed per unit weight of animal per unit time) or in (b) some behavioural activity such as the net spinning of trichopteran larvae.

The actual methodology required for recording and measuring such sub-lethal responses will vary and should be based upon laboratory experience of the chosen species. Clearly separate batches of animals must be used for mortality tests and sub-lethal tests. In order to obtain a time-related response the test containers should be examined as frequently as possible and certainly at 3 or 4 hourly intervals throughout the first 24 hours.

Data analysis

Lethal toxicity
For each batch of animals in a single container the cumulative percentage mortality occurring during the period of the test should be analysed. The sigmoid curve which generally results from a plot of cumulative percentage mortality against log time (Fig. 1) can be straightened for easier analysis by the "probit conversion" in which mortality is transformed to probability units by use of probit tables or more simply by plotting the data directly onto logarithmic probability paper. From the resulting probit line (Fig. 2) the median lethal time (LT50) can be determined. The slope function (S) which gives a measure of the pattern of response can be calculated from:

$$S = \left[\frac{LT84}{LT50} + \frac{LT50}{LT16}\right] X \frac{1}{2}$$

Confidence limits on both the LT50 and S can be derived using Litchfield's (1949) nomographic methods.

Sub-lethal toxicity
Specific methods for the analysis of sub-lethal toxic effects will depend upon the particular response determined but, if possible, some measure of the median effective time (ET50) should be arrived at using the same procedure as for lethal effects. Computer methods are available for determining LT50 and S values.

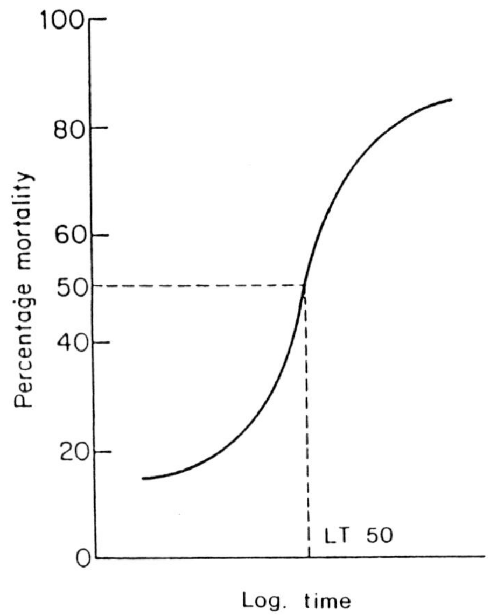

Fig. 1. The cumulative percentage mortality with time.

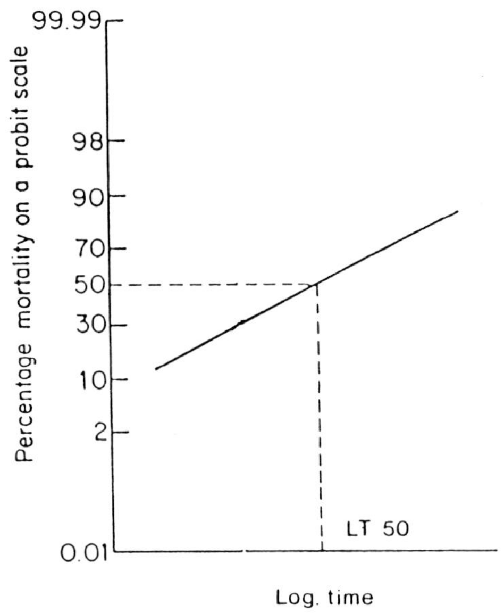

Fig. 2. The probit transformation, giving a linar plot, facilitates calculation of confidence limits for the LT50 and S.

Conclusions

These *in situ* bioassays will indicate the extent to which the effluent is toxic to the species present in the river and by comparing data recorded at various sites below the effluent the effectiveness of dilution in ameliorating toxicity can be seen.

Ideally these results should be interpreted together with data obtained from (a) routine biological surveillance; (b) *in situ* biomonitors such as the fish monitor; (c) laboratory toxicity tests; (d) chemical monitoring programmes.

References

Litchfield, J.T. (1949) A method for rapid graphic solution of time-percent effect curves. *J. Pharmac. exp. Ther.* 97:399-408.

Ormerod, S., Boole, P., McCahon, P., Weatherley, N., Pascoe, D. and Edwards, R. (1987) Short-term experimental acidification of a Welsh Stream: Biological effects of hydrogen ions and aluminium. *Freshwat. Biol.* 17:341-356.

Pascoe, D., Edwards, R. W. (1989) Single species Toxicity Tests. In: Boudou, A., Ribeyre, F. (eds) *Aquatic Ecotoxicology, Fundamental Concepts and Methodologies*, Vol. 11. CRC Press. 93-126.

Approaches to the Methodology of Using Ecosystem Components as Bioindicators of Radioactive Pollution

by D.A. Krivolutsky, A.D. Pokarzhevskij, and V.L. Usachev
Institute of Evolutionary Animal Morphology and Ecology
of the Russian Academy of Sciences, 11707 Moscow V-71, Leninsky Prospect, 33, Russia

Principle

To study the effects of ionizing radiation on terrestrial ecosystems it is essential to make observations on soil fauna. They comprise 90-95% of all animal species and 90-99% of total zoomass. Soil dwellers often prove to be an important factor in forest pest control, especially in radiation-affected stands. Terrestrial carnivorous arthropods are of special importance in the elimination of foci for animal infection on radionuclide-polluted territories, where they predate many parasites of vertebrates (especially mites), which spend a certain period of their life cycle in the soil. A close connection between the level of radioactive pollution and the ecology of agricultural pests in the soil is also observed. Their high population density is stable (amounting to many hundreds of thousands of individuals per 1 m^2 of the soil). Since they also have permanent feeding habits (phytophagous, saprophagous and carnivorous), soil arthropods present a convenient subject for elucidation of the ways and quantitative patterns of radionuclide migration in biogeocenoses, and may be used as bioindicators of polluted territories (Ghilarov and Krivolutsky, 1972).

Radionuclides, i.e. isotopes or chemical analogues of the basic biogenic elements which are readily incorporated into animal tissues and transmitted along food chains: ^{3}H, ^{14}C, ^{106}Ru, ^{131}I, ^{137}Cs, are of great importance in determining pathways. The following requirements are imposed upon animals used as bioindicators of industrial pollution by radioactive isotopes (Krivolutsky, 1983): (a) high density and metabolism level; (b) great length of life; (c) intensive reproduction; (d) a small home range; (e) sedentary way of life; (f) permanent contact with the anthropogenic factor under study; (g) available collection of mass samples; (h) sensitivity to the factor concerned; (i) fairly large size to facilitate dissection.

Fission products of uranium and plutonium, particularly long-lived ^{90}Sr and ^{137}Cs (radiochemical half-lives of 28.4 y and 30 y, respectively), have always attracted the attention of ecologists. These isotopes are found on the ground surface after nuclear weapon tests, as well as after the processing of the uranium-containing ores and reprocessing of nuclear fuel. Many radionuclides, including ^{90}Sr and ^{137}Cs, are rapidly assimilated in detectable amounts into the biogenic turnover and can be accumulated by animals. However, the radiation of these radionuclides is not high in comparison with natural radiation background and no deleterious effects are observed. Epigeous and soil-dwelling animals associated with the biogenic migration of elements, including man-made radionuclides, can serve as one of the biological indicators of environmental pollution with these toxic elements or substances. At the same time, the labelling of biological turnover in all living biospheric objects with man-made radionuclides opens up new possibilities and prospects for field ecology. This permits us to define flows of a substance in ecosystems more precisely, and to clarify trophic chains of organisms and their productivity.

Methods

The dosimetry of animals on plots with a high content of radionuclides

Radiometrical data of ^{90}Sr distribution patterns in forest cenosis, which are established by measuring the radionuclide concentration in different animal species included, make it possible to estimate radioactive doses, received both from external and internal radiation sources. Since ^{90}Sr and ^{90}Y beta-radiation have weak penetrating ability, most of beta-particle energy in a solid body is absorbed close to the origin of the radiation. So the pattern of the absorbed doses of beta-radiation virtually corresponds to that of ^{90}Sr in the soil profile.

Since the ^{90}Sr content in overground parts of the forest stand is much lower than in the soil, radiation doses of surface dwelling invertebrates proved to be much lower than doses to the fauna in soil. Doses in the soil, however, are not sufficiently high to produce a direct damaging effect of invertebrates, especially on their larval and imaginal stages. The decrease in population density may be possibly attributed to the fact that forest insects (including micro- and mesofauna) hibernate in the soil as eggs. This is the most radio-sensitive embryonic phase. In this case, irradiation dose is accumulated during several months and may be as great as 200-300 r.

In laboratory experiments (Tikhomirov, 1972), irradiation doses from several hundreds to several thousands r may cause the sterility of males and females in some species and egg destruction. In field conditions, irradiation damage is combined with that of some other adverse factors (e.g., low temperatures). This dual effect may cause animal destruction at much lower radiation doses compared with that under favourable conditions. It should be taken into account that the irradiation dose can be considerably increased (up to 50%) by internal irradiation, caused by incorporated ^{90}Sr. The authors showed cytogenetic effect of ^{90}Sr incorporated in *Crepis capillaris* seeds. The destructive effect of internal irradiation exceeds the external one more than tenfold. However, the absolute values of the incorporated irradiation doses in animals are not high; they are 2-3 orders lower than doses accumulated due to external radiation.

All the above provides conclusive dosimetric evidence for interpretation of the observed effects of irradiation on invertebrate populations in polluted forest ecosystems.

Earthworms are a particularly convenient subject for studies of the effect of an increased radiation background, as they are not only irradiated from the outside but also from the soil they swallow. All the other land animals under study do not swallow the soil and consume only animal or vegetative food in which the radionuclide content is considerably lower (by 1-2 orders) than in the soil.

The earthworms collected from the Radium V polluted soil showed considerable changes in the structure of integument epithelium (shape and size of mucous cells). The mucous cells of these earthworms are broader and larger as compared with controls. The number of mucous cells in the radiation-affected earthworms is much higher than that in the control ones. There are fewer cambial cells in these sections. The structure of the mid-gut epithelium was also studied. There are a large number of regeneration nests and individual regeneration cells in the basal part of the epithelium.

The effect of radioactive pollution on insects, which spend a certain period of their life cycle in litter, was studied by Tikhomirov (1973), who reported a predominant radiation effect on parasites.

As is seen, radioactive pollution exerts the greatest influence on the permanent soil dwellers by decreasing their population level, especially in deep soil layers, and by abrupt disturbances in the structure of animal communities. Temporary soil dwellers are possibly liable to some secondary effects (influence of heat on predators, parasites, fodder plants, etc.) rather than to the direct influence of radiation itself. They may show not only quantitative decrease, but even higher population density. In laboratory experiments with protozoa, their relatively high resistance was recorded; doses of about 100-150 Ci producing no marked effect. Doses of about 300 Ci revealed no influence in the experiments with soil *Amoeba terrestris*, excitation taking place even after treatment with a dose of 500 Ci. High resistance to ^{60}Co gamma-radiation was shown for Ciliata (*Colpoda maupasi, C. stenii*) from Uzbekistan serozems: 1-10 Ci activated their mobility and the rate of cell division, while 50-100 Ci depressed it. 300-500 Ci proved to be lethal for some of the organisms; survival rate depending on the dosage. Treatment with doses up to 100 Ci (at a rate up to 36000 R/min) caused the death of ciliates within 3-4 days.

Protozoological observations in control and ^{90}Sr polluted plots revealed differences in protozoan numbers and species composition. The survival of species depended on pollution fluctuations in microstacia, rather than on differences in the radioresistance of species (Korganova, 1973). The effect of radioactive pollution on Protozoa (one of the most resistant organisms) is to be explained by the high radiation level in water film, which consists of soil particles absorbed with ^{90}Sr.

Information on radiation-induced mortality in different species of soil animals is required to estimate ranges in radiosensitivity of species not examined. Such information would enable evaluation of possible animal problems in postnuclear attack environments, and would also be useful in control of invertebrates by irradiation.

In our investigations, 12 species of insects, 1 species of Enchytraeidae, 3 species of Diplopoda and 2 species of Lumbricidae were exposed to gamma radiation from cobalt-60 with total doses ranging from 10 to 200 kr. Invertebrates were irradiated and maintained under comparable conditions at 200°C. Mean and median life expectancy were estimated for each dose level. Correlations between sensitivity to radiation and taxonomic grouping were poor.

		LD 50/30 KR
Carabidae		
	Pterostichus oblongum punctatus	100
	Carabus violaceus	100
Staphylinidae		
	Astilbus canaliculatus	100
	Crephilus maxillosus	100
Scarabaeidae		
	Lethrus apterus	170
Tenebrionidae		
	Blaps lalophilum	210
	Blaps sp.	100
Diplopoda		
	Sarmatiulus kessleri	60-90
	Pachyiulus flavipes	100
	Cylindroiulus kuschkensis	100
Enchytraeidae		
	Enchytraeus albidus	120
Lumbriciae		
	Allolobophora caliginosa	70
	Lumbricus rubellus	80-160

Population radiation ecology of animals

It should be noted that the biological effects of ionizing radiation have been fairly well studied and described (see, for example, Coggle, 1983).

Population studies were carried out at levels of 0.1-10 Gr per year, i.e. in the dose-range for which biological effects are well studied, and which induce "radiation syndromes" in experiments on mammals. This was especially true for spermatogenesis and early embryogenesis, i.e., the most radiosensitive systems and stages in animal life cycles.

When conducting radioecological studies in natural ecosystems, the range of radiation doses for the natural communities in question is limited and depends on the concentration and migration of radionuclides in various components of the environment. The irradiation doses to animal populations are intimately correlated with the ecological mode of transport of radionuclides into the organisms, and the level of external radiation.

The studies on the root vole in plots with an elevated concentration of the natural radionuclides of U, Th and Ra have shown a number of morphophysiological defects and changes:

1. Lesions in the haemopoietic system and peripheral blood.
2. Delay in sexual maturation due to inhibition of spermato- and oogenesis.
3. Morphometric changes of organs (liver, spleen, testes, kidney).
4. Early death and resorption of up to 30% of the embryos.

5. A small increase in the frequency of chromosome aberrations in the bone marrow and its correlation with Ra content of the animal skull (Maslova and Materij, 1974).

Isaev (1975) revealed an increase in embryonic mortality by a factor of 1.5-4 in the rodents occurring in a ^{90}Sr-polluted plot at the radionuclide concentration exceeding 111 BK per g of skeleton. No reliable differences were found in the population density of rodents in the four experimental plots during seven years of continuous observations (Krivolutsky *et al.*, 1983).

The animals occurring in closed water bodies present a very convenient model for population studies of the effects of radiation, since the community is not replenished with outsiders.

The population structure changes due to the marked reduction in older age groups, which is equivalent to the reduction of life span. This is described in detail, for fishes (goldfish and roach) occurring in water bodies with a markedly elevated ^{90}Sr and ^{137}Cs content. In particular the fecundity decreases, the time of spawning is shifted, and the number of abnormal adults increases. All these changes took place at levels of radioactive pollution which excluded by three orders of magnitude the maximum permissible concentrations for water (Ermokhin and Muntyan, 1985).

Trophic level effects in biogenic migration of radionuclides

In spite of the abundance of literature, mainly radioecological, on the migration of elements in trophic chains, the patterns of the processes have not been studied enough to build a suitable universal model (Krivolutsky and Pokarzhevskij, 1977). According to the American radioecologists (Kitchings *et al.*, 1976), it is necessary to clear up the peculiarities of the transfer of an element from one trophic level to another one; element accumulation factors in organisms and the degree of retention of elements by organisms. The relationships between elements during migration in trophic chains is one of the most important and poorly studied factors. In experiments with laboratory and domestic animals, the problem of nonradioactive isotope effect on the radionuclides accumulation is considered in detail. However, for natural populations and ecosystems this problem is not solved, even for such well-studied pairs of elements as Ca-^{90}Sr and K-^{137}Cs.

Field studies on co-migration of these elements in terrestrial ecosystems are scarce (Sokolov *et al.*, 1987) and do not cover all the diversity of trophic chains of terrestrial animal communities.

Before estimating the co-migration of the elements in food cnains, let us pick out the fundamental features of elemental concentration in animal bodies (Pokarzhevskij, 1985). They are as follows: macroelement concentration in animal bodies doesn't depend on systematic inventory of animals and element content with the exception of Ca and S. Among animals there are Ca-concentrator groups: molluscs, diplopods, woodlice, and vertebrates to some extent. Calcium accumulation is a phenomenon which appeared in different groups of animals independently, and probably was connected with initial environmental conditions of ancestral forms. Sulphur behaves like microelements, as its concentration in animal bodies depends on its concentration in food.

Microelement (including Sr and Cs) concentration in animal bodies is determined by the species patterns on the one hand, and a microelement concentration in food on the other hand. The higher the microelement concentration in food, the higher their concentration in animal bodies. The dependency of the element content in bodies upon the element content in food increases with the increase of an element's atomic number as a whole.

The limits of an element's concentration in organisms are determined by its functional role in organisms and the possibilities for organisms to be viable at a given element concentration. But not all the chemical elements met in organisms play particular biological roles. Some of them are involved in turnover as "ecological impurities" along with element-carriers. At a certain level of substitution of element-carrier by the "ecological impurity", the toxic effect of the latter manifests itself. Sr and Cs seem to be the ecological impurities and therefore the concentration of the element-carrier in food must affect the Sr and Cs penetration into organisms.

The structure of a trophic chain has an important effect on the element transfer through it. Our last studies of element turnover and their migration through animal populations enabled us to propose a new conceptual model of a trophic chain which includes three heterotrophic levels: substrate-microorganisms - phytosubstrate [animals (phyto-consumers)] - animal substrate [animals (zooconsumers)]. The bifurcation of flows occurs for alkaline, alkaline-earth element and microelement migration in animal communities for saprovorous and herbivorous animals. One fraction of the element is assimilated in animal bodies from the substrate. The other penetrates through populations of microorganisms of the digestive tract. The important practical aspect of the flow bifurcation in the gastrointestinal tract is, that the more the element flows through microorganism populations the higher is its accumulation in animal bodies (Pokarzhevskij, 1987).

In consideration of the features of elemental concentration in animal bodies and food chain structure let us turn to the transfer of element pairs in food chains.

The observed ratio (OR) or the ratio of element pair concentration ratios in bodies to those in the diet are usually being considered from soil-plant links but we consider both soil and plants to be food substrates for microorganisms and animals.

The absence of data on element pair transfer in a substrate-microorganisms link and element flow bifurcation in gastrointestinal tracts forces us to consider OR in different animal groups to substrates. In invertebrates consuming both live and dead plant tissues, radionuclide concentration decreases compared with the element-carriers. This is due not only to the elective element accumulation but also to features of digestion, among them activity of microorganisms in the digestive tract. An increase in relative concentrations of Sr and Cs to element-carriers can be observed for carnivorous species. The consumption of particular animal tissues and accumulation of salts in insect excretion organs did not appear to matter.

Terrestrial and soil invertebrates comprise the diet of insectivorous animals. OR in relation to possible diet is less than 1 and OR is similar for insectivorous vertebrates from different classes.

It is well known from experiments that Sr concentration decreases relative to Ca when these elements are transferred from food to animals. However, the fluctuations of OR value in herbivorous vertebrates are noted. This is well demonstrated when comparing OR values for different rodent species, caught at various experimental plots, although large differences in values are not observed. The diet composition does not noticeably affect the OR values, either. In small mammals $OR_{Sr/Ca}$ is nearly 0.1 and $OR_{Cs/K}$ is about 0.5. In experiments with domestic animals $OR_{Cs/K}$ increase was seen to double; namely in poultry it is less than 1, in cows and goats it is nearly 1. Therefore there are no large inconsistencies in our data. As for carnivorous and omnivorous mammals their $OR_{Cs/K}$ is noticeably greater than 1. We demonstrated the same results for carnivorous reptiles and mammals. The concept of "trophic level effect" (Pendelton *et al.*, 1964), in which caesium concentration increases relative to potassium concentration at the successive trophic levels, appeared to be correct only for carnivorous vertebrates but not for a trophic chain as a whole. A consequent decrease in the Sr concentration relative to Ca is observed in a trophic chain including a prey-carnivore link.

So Sr seems to behave relative to Ca as "ecological impurity" and its relative concentration decreases as it migrates in a trophic chain. As for Cs, it is not as closely connected with K as Sr with Ca. In vertebrates with carnivorous or omnivorous food patterns, Cs transfer from diet to organism is more affected by digestive features than relationships with the element-carrier.

The distribution of these elements in biomass of particular components of an experimental ecosystem, corresponds to the ratio of element pairs at different trophic levels, but there are some deviations (Pokarzhevskij and Krivolutsky, 1975; Krivolutsky, 1983). Microarthropods accumulate Cs and Sr more intensively and earthworms less strongly than could be expected from K and Ca distribution in their biomass, and from their contents in substrates.

So despite existing correlations between Sr-Ca and Cs-K migration in trophic chains in terrestrial ecosystems, strict patterns of co-migration are not noted. This is due primarily to incomplete understanding of basic mechanisms of elemental migration in trophic chains.

Conclusions

The methods of radioecological research are very well described now, in the literature (Verkhovskaya, 1971; Schultz and Whicker, 1982). But explanation of field data from radioecological studies on animals, and use of wild animals as bioindicators of radioactive pollution, present some methodological difficulties. In our own investigations, it is shown that such species and population features as life cycle longevity, feeding patterns, population densities, connection with radioactive substrate, can have decisive importance for determining the ecological effect of ionizing radiation. At the ecosystem level it is clear that there is a significant role for microorganisms in understanding some effects in trophic chains. It is obvious that the success of practical bioindication is closely related to progress in such fields of fundamental ecology as population ecology and ecology of trophic chains.

References

Coggle, J. (1983) *Biological Effects of Radiation*. St. Bartolomew's Hospital Medical College, London.

Ermokhin, V., Muntyan S. (1985) A contribution to the estimation of the influence of an elevated level of ionizing radiation on the indices of animal populations. In: *Radioecology of Soil Animals*. Moscow, Nauka. pp. 148-156 (in Russian).

Ghilarov M.S., Krivolutsky D.A., 1972. Radioecological researches in soil zoology. *Proc. IV Coll. Pedobiol.*, Paris, pp. 537-544.

Isaev, S.I. (1975) Some questions of ecology of wild rodent reproduction in relation with their inhabiting of ^{90}Sr contaminated biogeocoenosis. *Ekologia* 1:45-51.

Kitchings, T., Digregoria, D., Van Voris P. (1976) A review of the ecological parameters in vertebrate food chains. In: Radioecology and energy resources. *Proc. 4th Nat. Symp. Radioecol.*, Corvallis (Oregon), 1975, Stroudsburg (Pa.). pp. 304-313.

Korganova, G.A. (1973) Influence of experimental soil pollution with ^{90}Sr on the fauna of soil protists. *Zool. J.* 52:939-941 (in Russian).

Krivolutsky D.A., Pokarzhevskij, A.D. (1977) The role of soil animals in nutrient cycling in forest and steppe. *Ecol. Bull.* 25:253-260.

Krivolutsky, D.A. (1983) *Radioecology of Terrestrial Animal Communities*. Energoatomizdat, Moscow.

Krivolutsky, D.A., Taskae, A.I., Testov, B.V. *et al.* (1983) Dynamics of animal population on plots with an elevated content of radionuclides in the soil. Trudy Komi filiala AN USSR. 90:1-40 (in Russian).

Maslova, K.I., Materij, L.D. (1974) Morphological changes in the peripheral blood and spleen of voles occurring in the environment with an elevated radioactivity. In: *Problems of Radioecology of Terrestrial Biogeocoenoses*. Syktyvkar. pp. 74-85 (in Russian).

Verkhovskaya, I.N. (Ed.) (1971) *Methods of Radioecological Research*. Atomizdat, Moscow.

Pendelton, R.C., Lloyd, R.D., Mays, C.W., Church, B.W. (1964) Trophic level effect on the accumulation of cesium in cougars feeding on mule deer. *Nature*. 204:708-709.

Pokarzhevskij, A.D., Krivolutsky, D.A. (1975) The role of pedobionts in biogeochemical cycles of calcium and strontium-90 in the ecosystem. In: *Progress in Soil Zoology*. Prague, Academia. pp. 249-254.

Pokarzhevskij, A.D. (1985) Geochemical ecology of terrestrial animals. Nauka, Moscow (in Russian).

Pokarzhevskij, A.D. (1987) A conceptual model of matter migration in soil animal communities. In: *Soil Fauna of Northern Europe*. Nauka, Moscow. pp. 34-38 (in Russian).

Schultz, V., Whicker, F.W. (1982) *Radioecological Techniques*. Plenum Press, N.Y., London.

Sokolov, V.E., Krivolutsky, D.A., Rjabtzev, I.A., Usatchev, V.L. (1987) Principles of using of wild animals as bioindicators of global radioactive pollution. *Acta Biol. Hung.* 38:59-68.

Tikhomirov, F.A. (1972) *The effect of Ionizing Radiation on Eecosystems*. Atomizdat, Moscow (in Russian).

Tikhomirov, F.A. (1973) The effect of gamma-ionization on the *Lymantria dispar* population. *Ekologia* 5:15-21 (in Russian).

The Utilization of Epigeic Invertebrate Communities as Bioindicators of Terrestrial Environmental Quality

by Vlastimil Ruzicka and Jaroslav Bohac
Institute of Landscape Ecology, Czech Academy of Sciences, Na sadkach 7,
37005 Ceske Budejovice, Czech Republic

Principle

Communities of diverse organisms have been used as bioindicators of the environmental status and particularly of human influence on ecosystems. A number of characteristics have been employed for the description of the structure of these communities (e.g. Schubert, 1985). It is clear that for practical utilization of the results of bioindication in landscape ecology several of the most sensitive characteristics with a large amplitude should be selected.

Data of selected characteristics should also be evaluated for a representative set of community samples from the area examined. This could form a basis for all successive comparisons. The degree of deterioration of the areas of interest then could be evaluated in absolute values, without having to resort to a comparison with some reference values with sporadic controls. This would actually be a gradual transition from bioindication to biodiagnostics. Various groups of epigeic arthropods, especially spiders, carabids and staphylinids, can be used in such biodiagnostic studies (Ruzicka, 1985, 1986; 1987; Bohac, 1986; Bohac and Ruzicka, 1988).

Methods

Methods of sampling of epigeic invertebrates

The epigeic invertebrates can be collected by the method of pitfall traps (Barber, 1931) or by the method of soil quadrat samples (Phillipson, 1971). It is very important for bioindicative studies to use one method for the comparison of sets from various biotopes. The invertebrates should be obtained by all-year sampling because the activity of some species is often restricted to a limited period of the year.

The use of indices of species diversity for bioindicative studies

The various indices of species diversity are calculated from the number of species and individuals in the sample. The value of indices of the species diversity can be used for the study of the human influence on ecosystems, because the high value of these indices is often characteristic for biotopes with stable environmental and biotic condition, diverse life condition, high stability and biological autoregulation. Communities with low value of species diversity are characteristic for ecosystems with strong human impact or for biotopes with extreme environmental conditions. For estimation of the human impact on communities of epigeic invertebrates we can use the following indices: the index of dominance (Simpson, 1949), the index of species richness (Margalef, 1958), the index of species diversity α (Williams, 1964), the Shannon-Weaver index (Shannon & Weaver,

1949) and the index of species diversity β (Odum, 1983).

On the other hand, the values of diversity indices based on the number of individuals and the number of species contains minimum information on species presented. They are dependent on many abiotic and biotic environmental factors. Diversity indices are suitable for common studies, for observing successive trends, for understanding principles of changes in communities. A suitable example of biotope evaluation by the use of index diversity α (Williams, 1964) is the study of staphylinid (Coleoptera, Staphylinidae) communities in various biotopes near Prague (Table 1). The value of the index α is calculated from the formula $S = \alpha \,.\, \log_e (1 + N/\alpha)$ (S = number of species, N = number of individuals). The accuracy of the index α verify the parameter x ($x = N/(n + \alpha)$) showing the statistical sufficiency of the data. The value of α is higher in communities sampled in plant associations in stage of paraclimax (*Carpinion betuli, Quercion robori-petrae, Alno-Padion*) than in secondary biotopes (dispersal belt, cultural forest) and in anthropogenous biotopes (ruderal biotope and field). Thus, diversity indices fail to suit as simple biodiagnostics indices of the degree of deterioration.

Table 1. The comparison of the species diversity α of staphylinid communities in various biotopes in Prague (after Bohac, 1986).

Biotope		N	S	α	x
Carpinion betuli	282	53	19.30	0.9359	
Quercion robori-petrae	176	41	16.77	0.9131	
Alno-Padion	451	81	28.74	0.9401	
Eu-Quercion pubescentis	112	16	5.09	0.9564	
Dispersal belt	239	40	13.74	0.9457	
Pino-Quercetum	167	23	7.33	0.9570	
Tilio-Acerion	45	9	3.36	0.9705	
Ruderal biotope	258	21	5.41	0.9795	
Field (wheat)	356	23	5.79	0.9848	

The characteristics of communities calculated on the base of qualitative ecological data (e.g. types of bionomy, life forms and ecological strategies) show more exactly the state of biotope and degree of human influence on it.

Percentage representation of species group in the communities as a biodiagnostic characteristic

In our study of the human influence on communities of epigeic invertebrates, the dividing of sampled material into groups was made easier following the relictness of occurrence (Buchar, 1983). Buchar distinguished the following species groups: relicts of the first rank are species living partly in relict habitats corresponding with the former climate types, partly in the original and natural forests, i.e. in the stands corresponding with the recent climate; and relicts of second rank are species living in the commercial forests and in other woody plant formations; with expansive species being able to form viable populations in artificially deforested habitats.

The term "relict" can bring protests from some zoologists and that is why we divided samples into three separate groups: I. Groups of protected territories (state nature reserves and unmanaged areas of protected landscape regions); II. Groups of territories not included in groups I and III; III. Group of managed territories (meadows and fields).

We divided collected species of arthropods in every sample in the above groups and calculated the percentage representation of number of species and specimens in individual groups (Table 2). The proportion of individual groups in the community show the degree of naturalness of the ecosystem. For example, in natural deciduous forest ecosystems the proportion of group III invertebrates is usually low, and that of groups I and II usually high. The optimal way to arrange samples from various biotopes is in ascending order, starting with the percentage representation of the number of specimens in the III group in the community at the bottom of the table (Fig.1). The frequency of the number of individuals of group III increases from protected biotopes with weak influence of man to biotopes with strong human influence (fields, ruderal biotopes, urban ecosystems, etc.). So, on the basis of the percentage representation of various individual groups in the community, we can determine the strength of the human impact on a biotope. This method of evaluation has many advantages over the use of mathematical indices of species diversity.

Table 2. Example of the calculation of the percentage representation of number of species and individuals in individual groups in the community of staphylinid beetles of beech forest.

Species	Group	No. of individuals
Atheta excisa (Epp.)	I	2
Domene scabricollis (Er.)	II	5
Mycetoporus brunneus (Marsh.)	II	1
Ocypus macrocephalus (Grav.)	II	1
Omalium rivulare (Payk.)	III	25
Othius punctulatus (Goeze)	II	4
Philonthus decorus (Grav.)	II	138
Philonthus fimentarius (Grav.)	III	1
Philonthus laevicollis (Lac.)	II	1
Quedium fumatus (Steph.)	II	1
Tachinus rufipes (L.)	III	7
All	11 species (100 %)	200 (100 %)
Group **I**	1 species (9 %)	2 (1 %)
Group **II**	7 species (64 %)	165 (83 %)
Group **III**	3 species (27 %)	33 (16 %)

The characteristics calculated from the relative frequency of the various groups of species in the community comply well with the sample arrangement with respect to the territory division (the number of inversions range from 0.38). The diversity indices correspond to the sample order by the territory groups rather poorly (the number of inversions range from 38-89). They are our knowledge on ecological demands of species evaluated in the calculation of percentual representation of species group in the community. The transition from rigorously quantitative to qualitative data is emphasized, e.g. by Fryer (1987). The best correspondence is found for the relative frequency of specimens of opportunistic species.

With the use of classification of relictness of occurrence, the situation at the sites examined can be characterized in absolute terms. In the context of the natural situation in Bohemia, the relative frequency of specimens of expansive species of spiders (Group III) in protected landscape regions attains a maximum value of 0.45, the relative frequency of specimens of relicts of the first rank (Group I) a minimum of 0.20, and the relative frequency of relicts of the first and second ranks (Group I and II) a minimum of 0.40 (Fig.1).

The correlation between the arrangement of ascending index of dominance and the ascending values of frequencies of expansive species of spiders (Fig.1) show that the order of index values follows roughly the sequence of the three territory groups but many exceptions exist, however. The most marked deviation of the characteristics from values typical of the respective territory group are more or less explicable. For most of the disorder, however, a feasible explanation is hard to find. For a community's characteristics to be a suitable biodiagnostic index of the degree of deterioration, it has to follow our division of the territories into the three groups very closely.

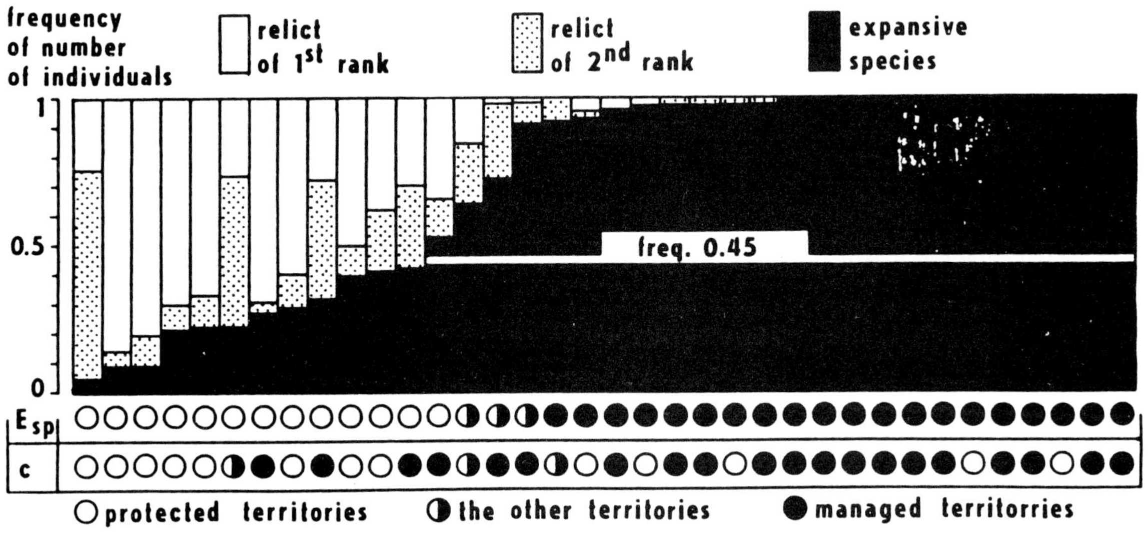

Fig.1. Bottom: Example arrangement of spider communities by ascending values of frequency of expansive species (E_{sp}) and of index of dominance (c = D2i). Top: Structure of relictness of species occurrence in the communities evaluated.

The relation between the frequency of the number of species and the frequency of the number of individuals of groups in the community: the use in bioindication

In order to find which relations exist between the two quantities taken into consideration, i.e., between the percentual representation of the number of specimens and of the number of species, each species group is plotted into graphs. Along the x-axis the number of specimans is plotted into percentage representation. This percentage representation has been calculated mainly from the values by order of magnitudes higher than the percentage representation of number of species and achieves a wider range of values. The percentage representation of the number of species we plot along the y-axis (Fig.2). Expressed numerically is the percentage representation of number of specimens accurately to tenths of a percent, the percentage representation of number of species accurately to the whole percent. The sum of the percentage representation of all species groups within one categorization must amount to 100%.

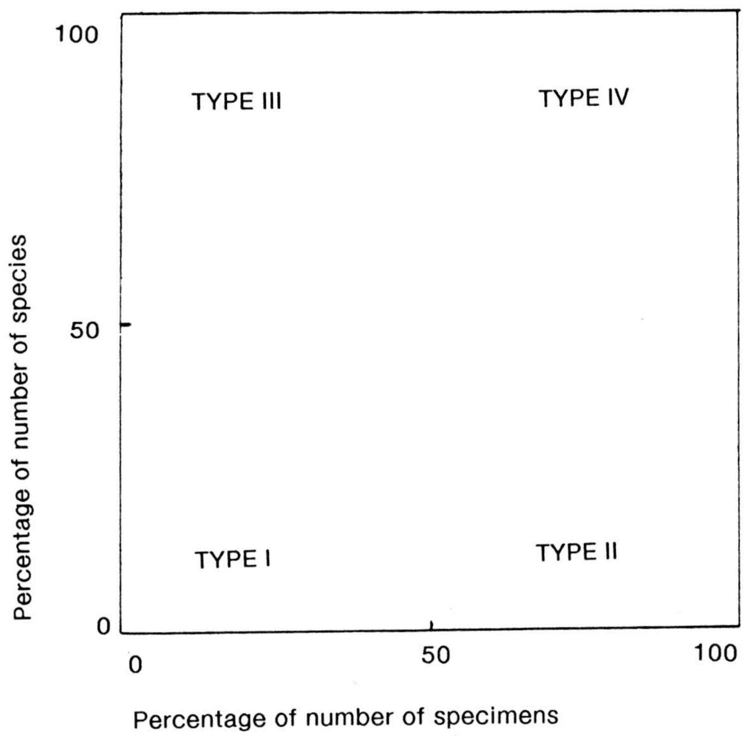

Fig.2. Basic diagram of the graph used. Illustrated are the position of four types of species groups with the extreme possibilities of the relation of the percentage representation of number of species.

By modelling, four limiting possibilities of the percentage representation of the number of specimens and of the percentage representation of the number of species can be traced out.

Species group of type I: The low percentage representation of the number of specimens and the low percentage representation of the number of species with the property under observation;

Species group of type II: The high percentage representation of the number of specimens and the low percentage representation of the number of species with the property under observation;

Species group of type III: The low percentage representation of the number of specimens and the high percentage representation of the number of species with the property under observation;

Species group of type IV: The high percentage representation of the number of specimens and the high percentage representation of the number of species with the property under observation.

These four limiting types of species groups we can preliminarily attach to four quadrants of the square graph (Fig.2).

In ecological classification the frequency of the number of species or the frequency of the number of individuals of a group in the community can be evaluated. In our classifications that reflect the ecological demands of species, some relationship exists between these two values for well-balanced communities; it can be portrayed by a logistic curve (Ruzicka, 1985) (Fig.3).

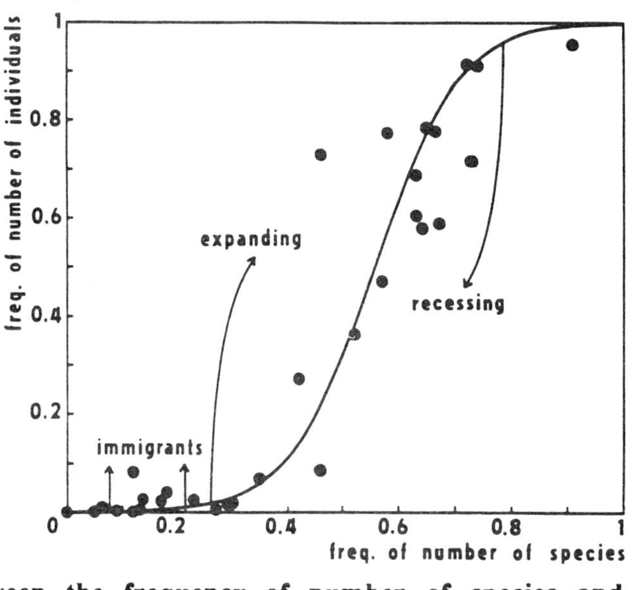

Fig.3. Relation between the frequency of number of species and the frequency of number of individuals for a set of relicts of first and second rank. Model of the shift of species groups is plotted.

The frequency of the number of individuals in a group of species gives information on the present state of the community and thereby, on the status of the abiotic environment concerned. In well-balanced communities the frequency of the number of individuals corresponds to the frequency of the number of species according to the logistic curve. In communities where such changes are just under way, this fragile balance can be disturbed.

A species group which increases in the frequency of the number of individuals, as compared to the equilibrium state, corresponds in their ecological demands to the newly establishing condition. Frequently, such a group is only represented by a few overmultiplied species. Species of a group whose frequency of the number of individuals is decreasing correspond in their ecological demands to the initial condition at the site. These may be the remaining members of previously more abundant populations. Components penetrating into the community from neighbouring ecosystems first attain values of frequency of the number of species which are considerably higher than the number of individuals (Fig.3).

If the parameters of the curves are obtained for various properties of species, the most important of which are the basic ecological factors, analysis of the situation in the community can indicate factors that are responsible for the changes which are just occurring in the community structure. Thus information can be gained about changes occurring at the site of interest via communities of organisms.

References

Barber, H.S. (1931) Traps for cave inhabiting insects. *Jour. Elish. Mitchell Sci. Soc.* 46:259-266.

Bohac, J. (1986) Kurzflügelkäfer (Coleoptera, Staphylinidae) als Bioindikatoren für Ökologisches Gleichgewicht einer Landschaft und menschlichen Einfluss. In: Paukert, J., Ruzicka, V., Bohac, J. (eds), *Proc. Fourth Int. Conf. Bioindicatores Deteriorisationis Regionis*, Liblice near Prague, Czechoslovakia, 1982. pp. 23-34.

Bohac, J., Ruzicka, V. (1988) The use of communities of epigeic invertebrates (Araneae, Coleoptera, Carabidae, Staphylinidae) for bioindication of the environmental quality. *Sov. Journ. Ecol.* 56-63.

Buchar, J. (1983) Die Klassifikation der Spinnenarten Bohmens als ein Hilfsmittel für die Bioindikation der Umwelt. *Fauna Bohemiae septentrionalis.* 1983:119-135.

Fryer, G. (1987) Quantitative and qualitative: numbers and reality in the study of living organisms. *Freshwater Biol.* 17:177-189.

Margalef, R. (1958) Information theory in ecology. *Gen.Syst.* 3:36-71.

Odum, E. (1983) *Basic Ecology.* Saunders, New York. 764 pp.

Phillipson, J. (1971) *Methods of Study in Quantitative Soil Ecology: Population, Production and Energy Flow.* IBP, Blackwell Sci. Publ., Oxford. 297 pp.

Ruzicka, V. (1985) Percentual representation of species group in the community (on example of spiders). *Ecology* (CSSR). 4:3-18.

Ruzicka, V. (1986) The structure of spider communities based upon the ecological strategy as the bioindicator of landscape deterioration. In: Paukert J., Ruzicka, V., Bohac, J. (eds), *Proc. Fourth Int. Conf. Bioindicatores Deteriorisationis Regionis*, Liblice near Prague, Czechoslovakia, 1982. pp. 217-237.

Ruzicka, V. (1987) Biodiagnostic evaluation of epigeic spider communities. *Ecology* (CSSR). 6:345-357.

Schubert, R. (1985) *Bioindikation in terrestrischen Ökosystemen.* VEB Gustav Fischer Verlag, Jena. 327 pp.

Shannon, G.E., Weaver, W. (1949) *The Mathematical Theory of Communication.* Univ. Illinois Press, Urbana. 323 pp.

Simpson, E.H. (1949) Measurement of diversity. *Nature.* 163:688.

Williams, C.B. (1964) *Patterns in the Balance of Nature.* Academic Press, London/New York. 415 pp.

An Approach to Biomonitoring the Industrial Pollution of the Landscape Using the Common Vole (*Microtus arvalis*)

by F. Sedlacek and V. Bejcek
Institute of Landscape Ecology, Czech Academy of Sciences, Na sadkach 7, 37005 Ceske Budejovice, Czech Republic

Principle

The common vole is an extremely abundant animal, even in strongly polluted landscapes. It thus represents an ideal test subject as a bioindicator of pollution exposure. The following physiological characteristics may be investigated: Heavy metal contents of hair; Weight of liver and spleen; Leucocyte assay (Leucogram); Haematocrit & haemoglobin content; Serum protein levels. These characteristics provide useful stress indices in comparing relative pollution status of landscapes.

Methods

Heavy metal contents of hair

Hair is known to accumulate toxicologically significant elements (e.g. Obrusnik, 1982). Elements of interest are determined by neutron activation analysis of back hair washed with acetone and redistilled water. Of 16 elements sought, elevated levels of 14-16 were found in industrially contaminated landscape. Concentrations of arsenic and scandium were one thousand per cent higher in an industrial site than in the reference background landscape (Obrusnik & Paukert, 1984).

Weight of liver and spleen

Changes in the size of the liver (the detoxifying organ) and the spleen (a key organ in the immunity system) are guide symptoms to many diseases. Weights of liver and spleen, expressed as a percentage of body weight could be a rough guide to exposure to industrial pollutants.

Data are presented in Fig. 1 which compare the distribution of spleen relative weights from the Trebon region - reference background landscape - with the industrial Most region. Liver relative weights exhibit a similar, but less pronounced, pattern (Bejcek *et al.*, 1982; Hejlkova *et al.*, 1983).

Leucocyte assay (Leucogram)

Total count and enumeration of leucocyte types, is one of the most basic investigations of the state of the immunity system. Conventional haematological methods were used to determine total count (Türk's solution, Bürker's chamber), and proportion of individual types (blood smear dyed according to Pappenheim). In the industrially exposed landscape - Most region - total leucocyte count was from 1.5- to 3-fold greater than in the background reference landscape. Data are presented in Fig. 2 for total count and count of types in the two landscapes over a three year period (Hejlkova *et al.*, 1983; Molikova *et al.*, 1985).

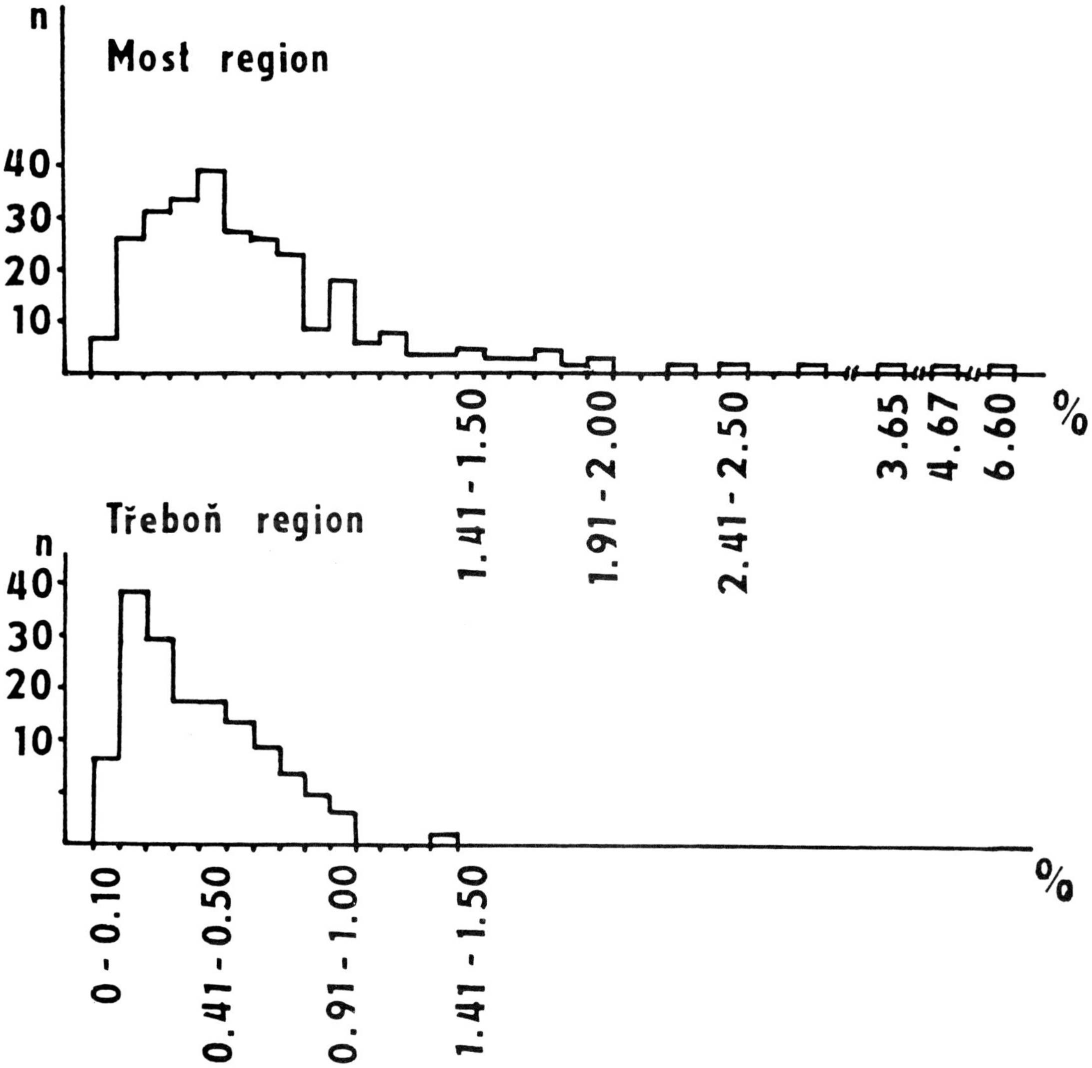

Fig. 1. Distribution of spleen relative weight in the common vole populations.

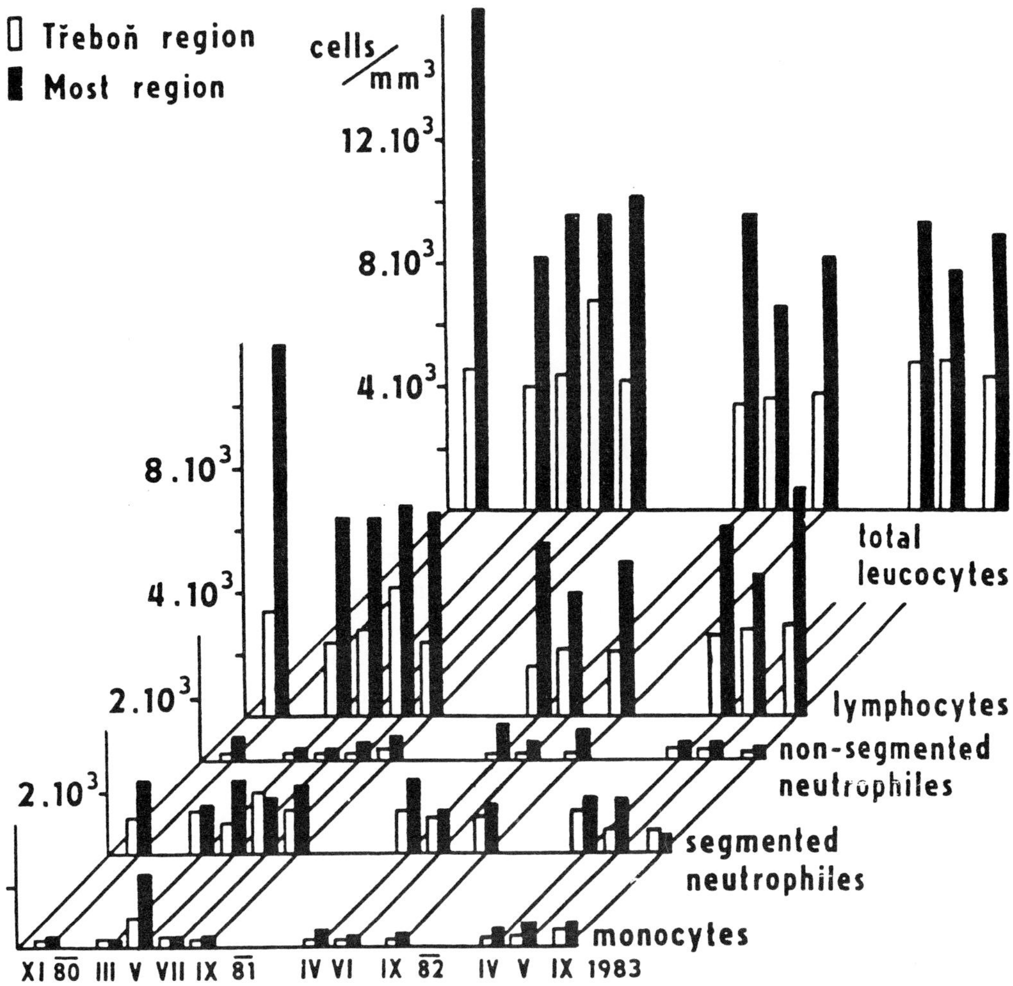

Fig. 2. Average leucocyte counts for the various capture terms.

Haematocrit and haemoglobin content

Production and development of red blood cells is a process that quickly reflects changes in internal and external environments. The red blood cell volume - haematocrit determined by a micromethod - was lower (45.3%) in the industrial landscape, than in the reference landscape (47.8%) (Sedlacek *et al.*, 1988). The concentration of haemoglobin was determined by the haemoglobin-cyanide method (Drabkin's solution, spectrophotometer). The concentration was lower (128 gr per litre) in the industrial landscape, than in the reference landscape (146 g per litre) (Sedlacek *et al.*, 1987).

Serum protein levels

Serum proteins have several functions, for example, transport, protection and nutrition. The development of electrophoretic methods has made it possible to study protein fractions as a function of different environmental factors.

Five fractions were determined by electrophoresis on cellulose acetate (25 x 145 mm foils ZA/MT and ZA/OT, Heintel Co. Austria). These were: albumin; alpha-1 globulin; alpha-2 globulin; beta globulin and gamma globulin. The levels of individual fractions in grams per litre, were obtained from the total protein content. The photometric Bio-lachema-test TP, CSSR, was used.

In industrially exposed landscape, higher levels of gamma-globulin and total protein were found in adult specimens. Subadult specimens born from March to June had higher levels of alpha-2 globulin.

References

Bejcek, V., Jirous, J., Stastny, K. (1982) Difference in the spleen weight of common vole (Microtus arvalis) between industrially exposed landscape and background landscape. In: *Environment of Country Landscape*. Proceedings of the conference. Brno, 1982, pp. 66-72 (in Czech).

Hejlkova, D., Jirous, J., Bejcek, V., Jecny, J. (1983) The impact of industrial emissions on the number of leucocytes and on the spleen and liver weight of common vole, *Microtus arvalis* (Pall.). *Ekologia* (CSSR). 2:75-84.

Molikova, D., Sedlacek, F., Bejcek, V., Stastny, K., Jirous, J. (1985) Changes in the leucograms of the common vole (*Microtus arvalis*) in an industrially polluted and in a background landscape. *Ekologia* (CSSR), 4:337-345.

Obrusnik, I. (1982) Use of activating analysis of hair by the control of environment pollution. *Chemicke Listy.* 76:805-825 (in Czech).

Obrusnik, I., Paukert, J. (1984) Indication of environmental pollution by means of INAA of the hair of some free-living mammals. *J. Radioanal. Nuclear. Chem.*, Articles. 83:397-406.

Sedlacek, F., Stastny, K., Bejcek, V. (1987) Concentration of haemoglobin in common vole (*Microtus arvalis*) from industrially exposed and background landscapes. *Proceedings of the Institute of Applied Ecology and Ecotechnics of the College of Agriculture, Prague.* 6:61-72 (in Czech).

Sedlacek, F., Stastny, K., Bejcek, V. (1988) Haematocrit value in common vole (Microtus arvalis) from industrially polluted and background landscapes. *Proceedings of District Museum in Most.* 7:55-59 (in Czech).

Comparative Biochemistry and Physiology

Effect of Detergents on Adrenergic Control of Blood Flow Through Isolated Perfused Fish Gills

By C.L. Bolis[1] and P. Canciglia[2]
(1) University of Milan, 9 via Balzaretti, 20133 Milan, Italy; (2) University of Messina, Sicily, Italy

Principle

Fish gills represent an important system for several physiological activities that control homeostatic processes. Gills have functions, other than respiratory, similar to those of the kidney in terrestrial vertebrates. Indeed, fish gills play an important role for ionic and osmotic balance, and for balance and excretion of nitrogenous substances. These different physiological activities are possible due to the anatomical structure of the gills such as large surface area, short diffusion distances, and countercurrent arrangements of water and blood flow.

For the exchange of water, oxygen and carbon dioxide, the process involved is diffusion (Isaia *et al.*, 1978; Loretz, 1979; Jackson and Fromm, 1981); all other processes require metabolic energy (Johansen and Pettersson, 1981)

The entire cardiac output of fish passes to the gills and then divides into two pathways: an arterio-venous pathway leading back to the heart and an arterio-arterial pathway leading to the systemic circulation (Laurent and Dunel, 1976). Several studies suggest that branchial blood flow in teleosts is controlled by circulating catecholamines (Wahlqvist, 1981; Bolis and Rankin, 1978). Catecholamines constrict the arterio-venous pathway and dilate the arterio-arterial pathway (Payan and Girard, 1977).

Autonomic nerves are also involved in the regulation of gill blood flow, adrenergic fibres constrict the arterio-venous and dilate the arterio-arterial pathway while cholinergic fibres constrict the latter (Nilsson and Pettersson, 1981).

Teleost and elasmobranch gills have the same basic geometry and both operate as countercurrent exchangers. The gill is composed of three structural subunits: the gill arches supporting the filaments which support the lamellae. The filaments are branched in parallel on the gill arch, and their vascular circulation comprises two arteries (afferent and efferent) and a vein.

In the filament there are two distinct vascular compartments: one located between the afferent and the efferent artery - the lamellar compartment; the central venous sinus forms a separate vascular compartment inside the filament. Thus it forms an alternate pathway for the blood flowing through the lamella which, instead of taking the arterial route to the dorsal aorta, may be recycled via the heart into the branchial circulation (Payan *et al.*, 1984).

The gill constitutes the principal surface of contact with the environment and, in view of its particular role in the control of several homeostatic processes, any damage of the structure may induce the damage of functions, either based on diffusion processes or, even worse, on functions relying on high metabolic energy demand.

Detergents are molecules which decrease the surface tension of structures. Detergent molecules are comprised of a hydrophobic portion with a long hydrocarbon chain (C12-C18) and a hydrophilic portion which can be a positive ion (cationic detergent), a negative ion (anionic detergent), or a non-ionized group polar enough to give rise to relevant attractive forces with water molecules (non-ionic detergents). Since teleost gills expose a large surface area to the environment, pollutants present in water come into close contact with the branchial epithelium (Abel and Skidmore, 1975; Abel, 1976) where they perform their major toxic action. Among detergents, the most studied for toxicity in aquatic animals is LAS (linear alkylbenzensulphonate). Water salinity increases detergent toxicity.

Data are consistent with the action of detergent (LAS) on β-adrenergic receptors situated between afferent and efferent branchial arteries, playing a vital role in the control of blood flow through teleost gills (Rankin and Maetz, 1971; Wood, 1974; Stagg *et al.*, 1981). β-adrenergic receptors are located within the cell membrane together with the adenylate cyclase subunits with which they interact (Ori and Schramm, 1976; Limbird and Lefkowitz, 1977). The lipid constituents of the plasma membrane consist of amphipathic units resembling detergent molecules and dissolution of the detergent in the membrane is to be expected. At high concentration, this results in the formation of mixed micelles and solubilization of the membrane (Helenius and Simons, 1975).

Low concentrations may increase membrane fluidity and affect β-receptor adenylate cyclase interactions. There is also evidence of high-affinity binding of detergent to protein molecules, including adenylatecyclase (Neer, 1978). β-adrenergic mechanisms producing increased cardiac output and decreased branchial vascular resistance are of great importance to fish, especially during exercise (Bolis and Rankin, 1975).

Methods (Rankin and Maetz, 1971)

Preparation of gills for perfusion

An intraperitoneal injection of 5000 i.u. heparin is given half an hour before sacrifice. The fish was then weighed, and killed by decapitation just posterior to the operculum. The gills were exposed by cutting from the operculum to the apex of the lower jaw on both sides with scissors. The gills and associated muscles could then be lifted with forceps, and the gills cut free from the roof of the pharynx with small blunt-tipped scissors. The gill arches were carefully separated, and the arch to be cannulated was retained while the others were placed in a beaker of heparinized Ringer solution and stored at 4°C for cannulation later.

The afferent and efferent arteries were easily visible as long as they contained blood, and the afferent vessel could be cannulated, usually without further dissection, with a short length of polythene tubing (no. 5; i.d. 1.19 mm; e.d. 1.70 mm) drawn out at the end, the other being attached to a syringe containing 2 ml heparinized Ringer solution. The efferent artery was cleared back to a point where a subsidiary artery branched off. This bifurcation proved a convenient point at which to make a small cut through which the efferent cannula could be inserted. As soon as the first ligatures had been tied to hold the cannulae in place, Ringer solution was injected through the afferent cannula to clear the gill of blood. Further ligatures were then tied around the cut ends of the gills to include the cannulae and blood vessels; these helped to hold the cannulae in position and to prevent leakage of perfusate from minor vessels.

Perfusion

The gill was suspended by the cannulae in a bath of aerated sea-water or fresh water maintained at 15.5 ± 0.5°C. The two cannulae were fitted tightly into cannulae of larger diameter filled with Ringer solution, taking great care not to let any air bubbles enter the gill. In some experiments two gills were mounted side by side, one serving as a control. The gills were perfused under hydrostatic pressure from a reservoir at a variable height (usually 40-50 cm above the bath water level). The outlet cannula terminated in an electronic drop counter which was also at a variable height (usually 20-30 cm) above the bath water level. The output of the drop counter was recorded on a potentiometric chart recorder. The inlet and outlet pressures were thus similar to the ventral and dorsal aortic blood pressures of the fish *in vivo*.

Composition of Ringer solution

Fish Ringer solution was prepared with an ionic composition based on that of fish plasma. The Ringer solution used for freshwater fish differed slightly from that for seawater fish, different quantities of stock solution being used to make up the final Ringer solution (Table 1). This Ringer solution was found to give better results than Krebs-Ringer solution as far as the maintenance of a good perfusion rate was concerned. It was found that reducing the calcium concentration to 0.5 mmol/1 helped to prevent spontaneous vasoconstriction, and this concentration was used for most experiments. In some experiments the magnesium concentration was increased to 10 mmol/1. The Ringer solution was gassed with 95% O_2; 5% CO_2 throughout the experiments, giving a pH of 7.35. The Ringer solution was filtered through a 0.22 μm Millipore filter before use.

Concentration-response curves were obtained by increasing the concentration of test substance by a factor of ten as soon as the flow had stabilized in the presence of the previous concentration, i.e., they were cumulative logarithmic concentration-response curves. Flow rates were expressed as percentages of control flow.

Table 1. Composition of Ringer solution

	Concentration of Stock Solutions (g/l)	ml stock soln/l Ringer soln		Final concentration mmol/l	
		F.W.R.	S.W.R.	F.W.R.	S.W.R.
NaCl	167.2	40	50	114.4	143.1
$NaHCO_3$	44.0	50	50	26.2	26.2
$NaHPO_4$. $12H_2O$	29.0	20	20	1.6	1.6
KCl	12.5	12	20	2.0	3.4
KH_2PO_4	2.5	20	20	0.3	0.3
$CaCl_2$	14.0	10	10	1.3	1.3
$MGSO_4.7H_2O$	50.0	10	20	2.0	4.1
$(NH_4)_2SO_4$	5.0	10	10	0.4	0.4

20 g polyvinyl pyrrolidone and 1 g of glucose were added per litre Ringer solution.
F.W.R. = Freshwater Ringer solution; S.W.R. : Seawater Ringer solution.

Fish gills respond very promptly to the action of noradrenaline, the trout gill being more sensitive than eel gill. As an example, see Figs 1-4.

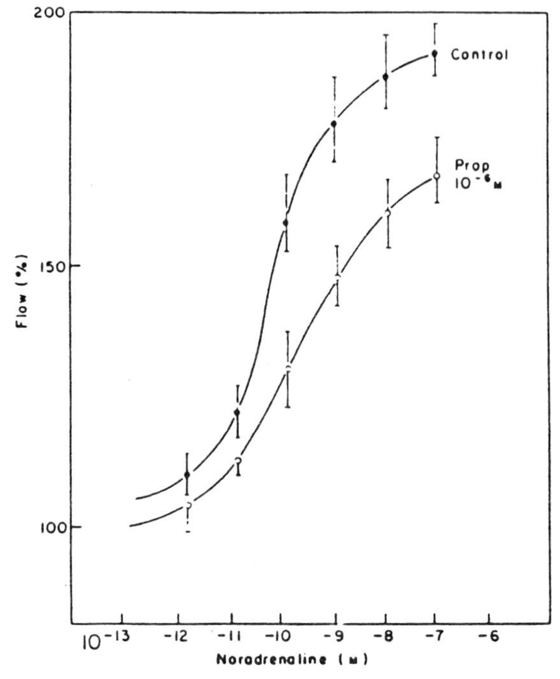

Fig. 1. Concentration-response curves for vasodilatory action of noradrenaline on eel gills in the absence (solid circles) and presence (open circles) of 10^{-6}M propranolol. Ordinate; flow as % of control flow. Abscissa; molar concentration of noradrenaline. Vertical bars represent ±1 S.E.M. (n=15 in all cases).

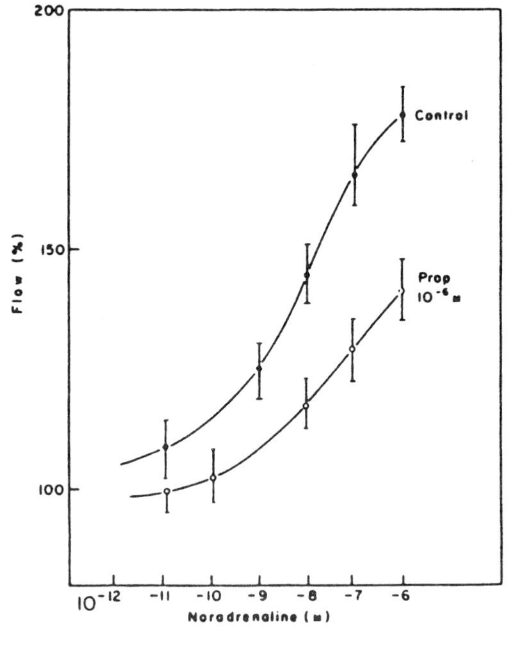

Fig. 2. Concentration-response curves for vasodilatory action of noradrenaline on trout gills in the absence (solid circles) and presence (open circles) of 10^{-6}M propranolol. Ordinate; flow as % of control flow. Abscissa; molar concentration of noradrenaline. Vertical bars represent ±1 S.E.M. (n=15 in all cases).

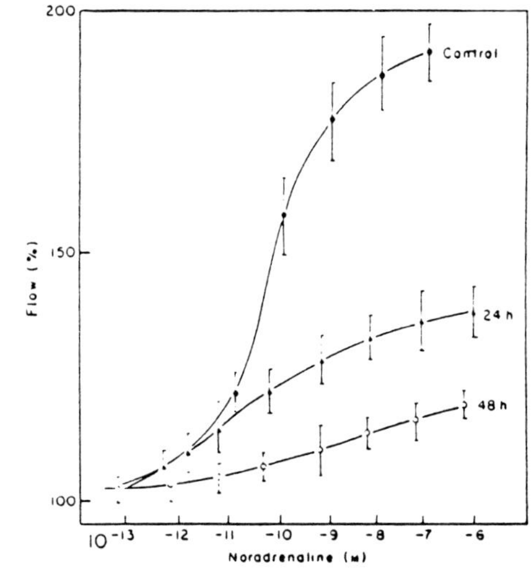

Fig. 3. Concentration-response curve for vasodilatory action of noradrenaline on control eel gills (closed circles) and on gills of eels kept for 24h (triangles) or 48 h (open circles) in 1 mg l^{-1} LAS. Ordinate; flow as % of control flow. Abscissa; molar concentration of noradrenaline. Vertical bars represent ±1 S.E.M. (n=15 in all cases).

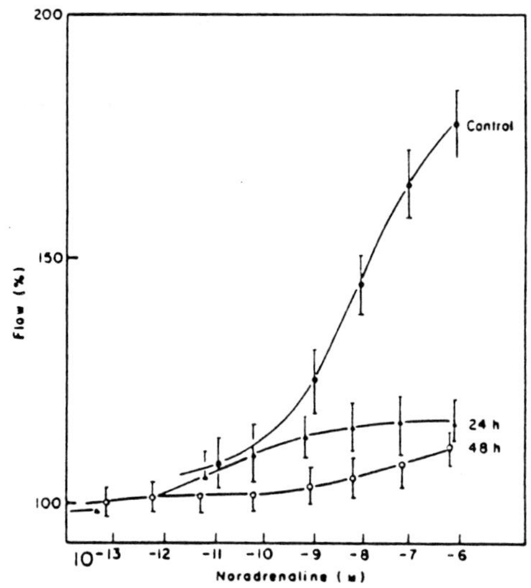

Fig. 4. Concentration-response curve for vasodilatory action of noradrenaline on control trout gills (closed circles) and on gills of trout kept for 24h (triangles) or 48 h (open circles) in 1 mg l^{-1} LAS. Ordinate; flow as % of control flow. Abscissa; molar concentration of noradrenaline. Vertical bars represent ±1 S.E.M. (n=15 in all cases).

In performing this method, the following should be taken into consideration:

1. catheterization of the afferent and efferent branchial arch arteries should be performed as fast as possible in order to prevent branchial ischemia;
2. careful control of potential leakage of perfusate into the external medium; and
3. control of mucus accumulation on gill filaments should be performed.

References

Abel, P.D. (1976) Toxic action of several lethal concentrations of an anionic detergent on the gills of brown trout (*Salmo trutta* L.) *J. Fish Biol.* 9:441-446.

Abel, P.D., Skidmore, J.F. (1975) Toxic effects of an anionic detergent on the gills of rainbow trout.

Wat. Res. 9:759-765.

Bolis, L., Rankin, J.C. (1975) Adrenergic control of blood flow through fish gills: Environmental implication. In: *Comparative Physiology - Functional Aspects of Structural Materials*, Bolis, L., Maddrell, S.H.P., Schmidt-Nielsen, K. (eds). Amsterdam, North-Holland Publishing Co. pp. 223-233.

Bolis, L., Rankin, J.C. (1978) Vascular effects of acetylcholine, catecholamines and detergents on isolated perfused gills of pink salmon, *Oncorhynchus gorbusha*, coho salmon, *O. kisutch*, and chum salmon, *O. keta*. *J. Fish Biol.* 13:543-547.

Bolis, L., Rankin, J.C. (1980) Interactions between vascular actions of detergent and catecholamines in perfused gills of European eel, *Anguilla anguilla* L., and brown trout, *Salmo trutta* L. *J. Fish Biol.* 16:61-74.

Ganmo, A., Kollberg, S. (1976) Uptake pathways and elimination of a nonionic surfactant in cod (*Gadus morhua* L.). *Wat. Res.* 10:189-194.

Helinius, A., Simons, K. (1975) Solubilization of membranes by detergents. *Biochim. Biophys. Acta.* 415:29-79.

Hughes, G.M. (1966) The dimensions of fish gills in relation to their function. *J. Exp. Biol.* 45:177-195.

Hughes, G.M. (1976) Polluted fish respiratory physiology. In: *Effects of Pollutants on Aquatic Organisms*. Lockwood, A.P.M. (ed.) Cambridge University Press. p. 163-183.

Isaia, J., Maetz, J., Haywood, G.P. (1978) Effects of epinephrine on branchial nonelectrolyte permeability in rainbow trout. *J. Exp. Biol.* 74:227-237.

Jackson, W.F., Fromm, P.O. (1981) Factors affecting ^3H_2O transfer capacity of isolated perfused trout gills. *Am. J. Physiol.* 240:R235-R245.

Johansen, K., Pettersson, K. (1981) Gill O_2 consumption in a teleost fish, *Gadus morhua*. *Resp. Physiol.* 44:277-284.

Laurent, P., Dunel, S. (1976) Functional organization of the teleost gill, 1. Blood pathways. *Acta Zool.* 57:189-209.

Limbird, L.E., Lefkowitz, R.J. (1977) Resolution of adrenergic receptor binding and adenylate cyclase activity by gel exclusion chromatography. *J. Biol. Chem.* 252:799-802.

Loretz, C.A. (1979) Water exchange across fish gills: The significance of tritiated-water flux measurements. *J. Exp. Biol.* 79:147-162.

Neer, E.J. (1978) Size and binding of adenylate cyclase from bovine cerebral cortex. *J. Biol. Chem.* 253:1498-1502.

Nilsson, S., Pettersson, K. (1981) Sympathetic nervous control of blood flow in the gills of the Atlantic cod, *Gadus morhua*. *J. Comp. Physiol.* 144:157-163.

Ori, Y.J., Schramm, M. (1976) Coupling of catecholamine receptor from one cell with adenylate cyclase from another cell by cell fusion. *Proc. Natl. Acad. Sci.* 73:4410-4414.

Payan, P., Girard, J.-P. (1977) Adrenergic receptors regulating patterns of blood flow through the gills of trout. *Am. J. Physiol.* 232:H18-H23.

Payan, P., Girard, J.-P., Mayer-Gostan, N. (1984) Branchial ion movements in teleosts: the roles of respiratory and chloride cells. In: *Fish Physiology. vol. X, Gills, Part B, Ion and Water Transfer*, Hoar, W.S., Randall, D.J. (eds). Academic Press, Inc.

Rankin, J.C., Maetz, J. (1971) A perfused teleostean gill preparation: vascular actions of neurohypophysial hormones and catecholamines. *J. Endocr.* 51:621-635.

Stagg, R.M., Rankin, J.C., Bolis, L. (1981) Effect of detergent on vascular responses to noradrenaline in isolated perfused gills of the eel, *Anguilla anguilla* L. *Env. Pollution* (Sec. A). 24:31-37.

Wahlqvist, I. (1981) Brancial vascular effects of catecholamines released from the head kidney of the Atlantic cod, *Gadus morhuza*. *Mol. Physiol.* 1:235-241.

Wood, C.M. (1974) A critical examination of the physical and adrenergic factors affecting the blood flow through the gills of rainbow trout. *J. Exp. Biol.* 60:241-265.

Quantitative Assessment of Gill Damage Using Relative Morphometric Diffusing Capacity

by G.M. Hughes
Research Unit for Comparative Animal Respiration, University of Bristol, Bristol BS8 1UG, U.K.

Principle

The rate of transfer of gases across respiratory surfaces (V) is determined by the area (A) and thickness (t) of the barrier as well as a constant (K) for the particular material which constitutes that barrier.

For oxygen: $VO_2 = A/t.K.\Delta PO_2$, where ΔPO_2 is the mean difference in oxygen tension across the barrier. The ratio $VO_2/\Delta PO_2$ is the physiological diffusing capacity (DO_2) and $A/t.K$ = morphometric DO_2.

In aquatic animals gill area and water/blood distance are the relevant measurements but unfortunately few measurements are available for the permeation co-efficients (K) for oxygen and carbon dioxide - usually values based on Krogh's measurement for frog connective tissue are used. Adverse environmental conditions may affect both the area and thickness of the respiratory surface and thus interfere with gas exchange.

Morphometric methods have been developed (Hughes and Perry, 1976) which make it possible to measure such changes and hence to gain a direct indication of the extent to which oxygen uptake can be reduced as a consequence of environmental pollution. Although external as well as microscopic observations of gills have been used as indicators of environmental conditions, the methods described here provide an objective and quantitative index for these effects.

Methodology

Although methods are available for direct determination of the gill area of fish (Hughes, 1984), such measurements are not essential for the present purpose. Thus diffusing capacity (DO_2) of the gills of fish from unpolluted waters = $A_{norm}/t_{norm}.K$. The relative change in diffusing capacity of gills of the same species from a polluted environment is given by the relative diffusing capacity:

$$Drel = A_{pol}/A_{norm} \times t_{norm}/t_{pol}$$

(Assuming the same permeation co-efficient for the gills of both groups of fish.)

It is therefore necessary to obtain estimates for values which are proportional to the ratio A/t for each group of fish in order to determine Drel. This can be achieved using high power light microscopy and semi-thin sections (1 μm).

Preparation of Material

The fish is killed and portions of each gill arch are removed from one side of the fish. For small fish these would be whole filaments but for larger specimens only filament tips may be removed using fine scissors with curved blades. The filaments are placed in 2.5% glutaraldehyde fixative buffered at pH 7.6 and in a carrier medium of appropriate osmolarity. For smaller fish, groups of two or three filaments attached together form ideal material which is post-fixed in 1% osmium tetroxide in the same buffer/carrier and embedded in resin (Epon or Araldite) using standard procedures.

Sectioning is carried out at a thickness of about 1 μm using an ultramicrotome. Orientation of the material during sectioning is very important, as the method requires the plane of the section be perpendicular to the main surface of the secondary lamellae (Fig. 1). From the diagram it is apparent that if the section passes at right angles to the lamellar surfaces the number of lamellae per mm will be the same as that which will have been counted before sectioning by direct observation of the material within the resin. However, if the plane of the section deviates, the number of lamellae per mm will decrease. Such sections are rejected if the count differs by 10% of the original value. The sections are stained in methylene blue and mounted according to standard EM procedures.

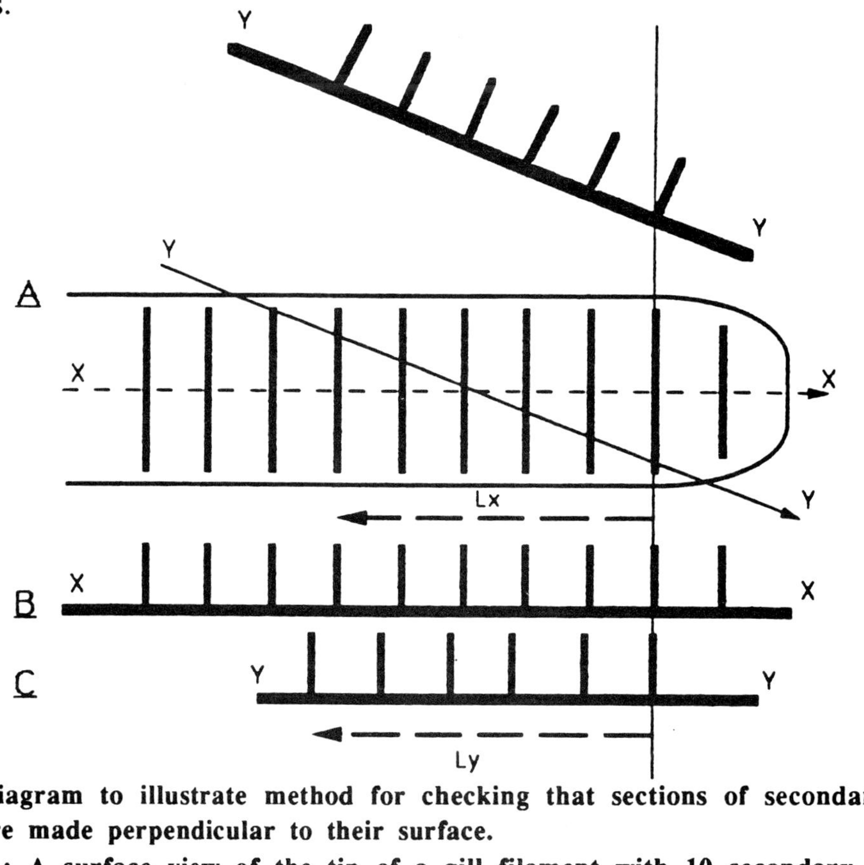

Fig. 1. Diagram to illustrate method for checking that sections of secondary lamellae are made perpendicular to their surface.
Diagram A: A surface view of the tip of a gill filament with 10 secondary lamellae: section x - x is the required perpendicular section and the frequency of the secondary lamellae is shown below.
Diagram B: A nonperpendicular section (y - y) produces the appearance shown (C) and it is apparent that the distance for five lamellae (Ly) is significantly greater than for the perpendicular section (Lx).

Morphometric Examination of Sections of Gill Lamellae

A good high-powered microscope with oil immersion objective is required so that material can be viewed at magnifications of at least 1000x. In an ideal set-up the image is observed on a projection screen containing a morphometric grid. A Merz grid (Fig. 2) is appropriate. If a suitable microscope is not available then high magnification photomicrographs can be used but as a large number of fields must be sampled this would be expensive. Alternatively 35 mm negatives can be projected on to a suitable grid. Another possibility is to have a transparency of the grid mounted within the eyepiece of a microscope but such procedures are not easy. Other equipment facilitating morphometry is some counting device. This method requires (a) counting the number of intersections between the sinusoidal test lines and the outer (Io) and inner (Ic) surfaces of the water/blood barrier; (b) measuring the distance between intersections with the outer surface and the nearest red blood cell. This latter measurement is easily achieved with a projection screen or photomicrographs but is more difficult to carry out using an eyepiece micrometer.

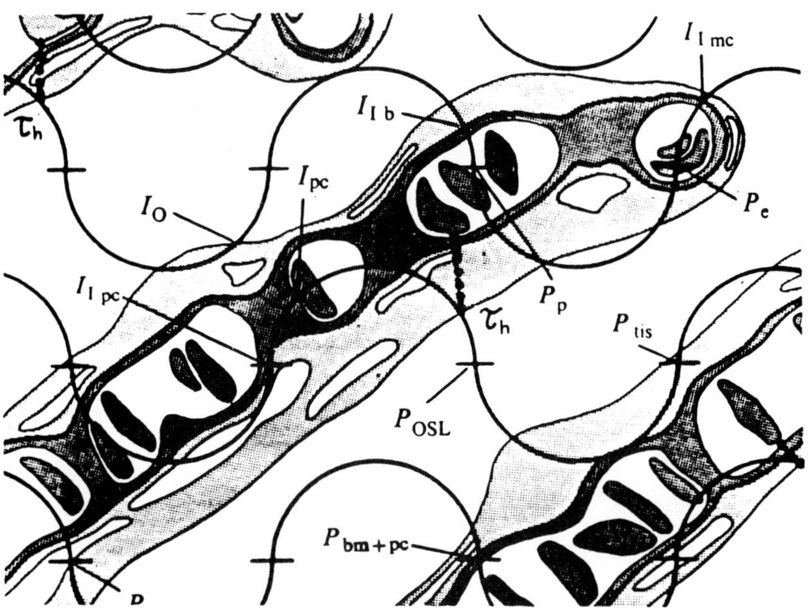

Fig. 2. Portion of Merz grid superimposed on diagrammatic section through secondary lamellae illustrating points (P) and intersections (I) counted. Distances (τ_h) from intersections with outer surface to nearest red blood cell were used to compute harmonic barrier thickness.

The appropriate measurement of thickness is the harmonic mean. Its ease of measurement and accuracy is assisted by a suitable ruler in which the shorter distances are measured more accurately than the larger ones (Fig. 3). Such a procedure is necessary as the contribution of each measurement is proportional to the reciprocal of the distance. In practice the measurements are classified into 10 groups (or perhaps 13 in Fig. 4) and it is only necessary to decide which of the 10 sections of the rule the particular distance falls.

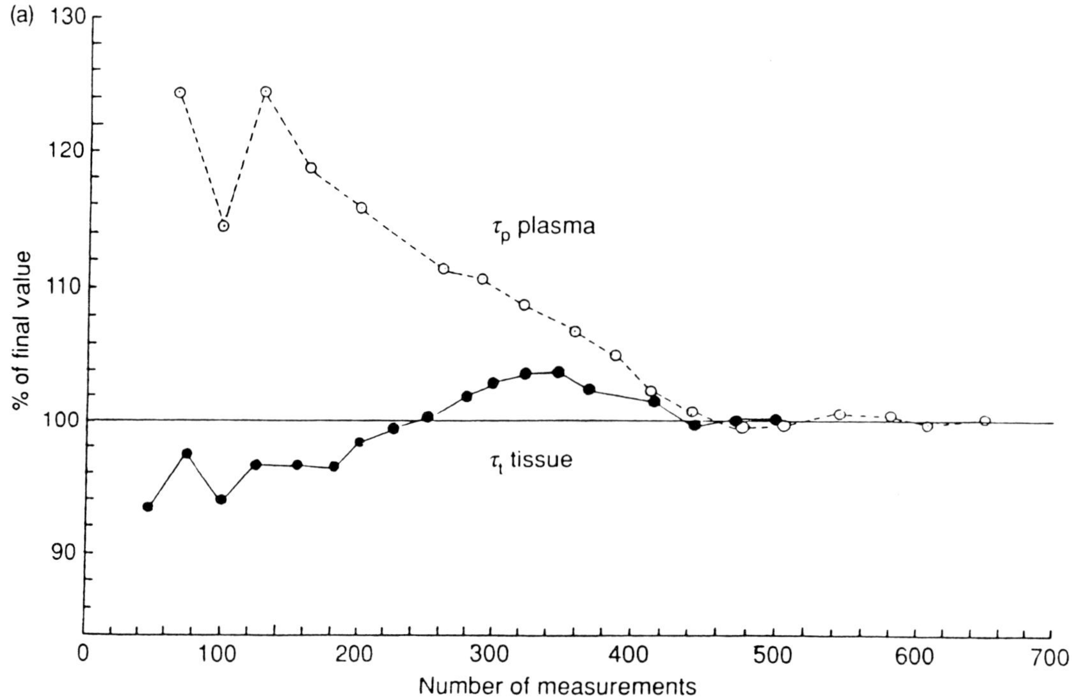

Fig. 3. Plots showing change in estimate of harmonic mean thickness of the tissue (τ_t) and plasma layers (τ_p) of the water/blood barrier of a fish secondary lamella. Values are plotted as a percentage of the final result (from Hughes *et al.*, 1992).

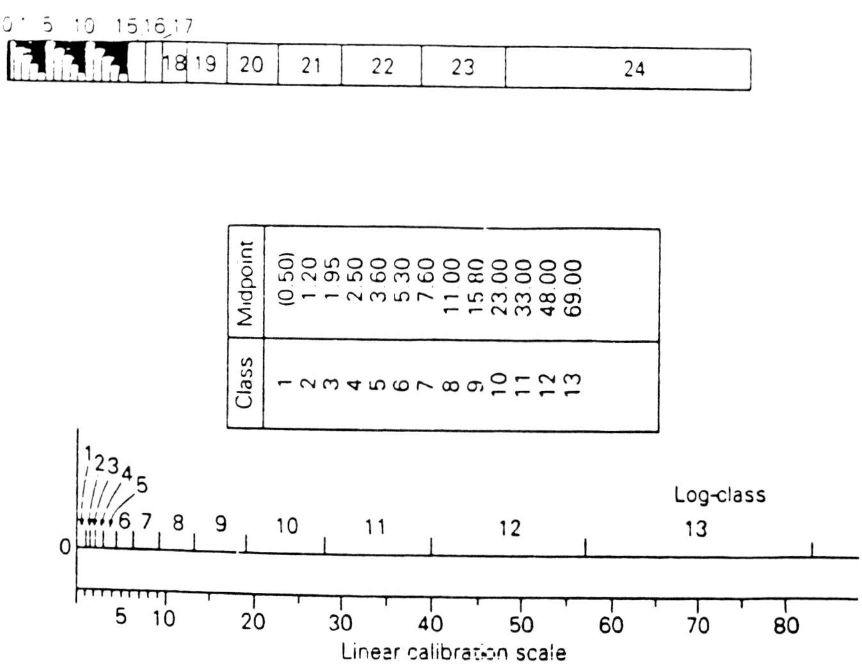

Fig. 4. Examples of rule used to measure harmonic mean thickness of water/blood barrier of secondary lamellae (from Hughes, 1989).

These measurements are then inserted in the following relationship to give the harmonic mean.

$$N/(n_1/L_1 + n_2/L_2 \ldots\ldots\ldots\ldots + n_{10}/L_{10}) \times 2/3 = \tau h \quad \text{(harmonic mean distance)}$$

where n_1 - n_{10} are the number of occurrences of the measurement using a scale divided into 10 sections in which the mean values (L_1 - L_{10}) represent the distances for the midpoint for each of the ten sections, N = total number of measurements.

For more detailed accounts of this and other morphometric methods consult standard works on stereology (e.g., Weibel, 1979).

An important aspect of all morphometric procedures is sample size. One advantage of a light microscopic method is the possibility of sampling from filaments from each of the gill arches. Perhaps three blocks should be made from each of the five arches and then a random choice made from one of each of them. From each of these five blocks sections will be obtained and measurements carried out on a sufficient number of filaments to ensure that the total number of measurements for each parameter is large enough to achieve some constancy. For example, Fig. 3 shows the cumulative result of an estimate for barrier thickness in which the value gradually approaches the final mean value obtained using a large number of sections. From such a test run it was decided that only 500 measurements were needed to achieve a value close to that which would have been obtained when a much larger number of measurements were made. These procedures are important not only in order to achieve statistically valid measurements but also because they reduce the total effort required, as morphometry is very time consuming.

Detailed Calculation of Drel

As indicated above the counts are made of intersections between the grid lines and the inner and outer borders of the barrier and these values are each proportional to the surface area of their respective boundaries. Consequently the relationship for Drel now becomes:

$$I_{o,pol}/I_{o,non} \times t_{non}/t_{pol} = \text{Drel}$$

Additional morphometric data can be obtained if counts are made of other features of the secondary lamella as described in Hughes and Perry (1976). A particularly useful one is to determine the proportion of the barrier which is occupied by non-tissue spaces, sometimes referred to as lymphoid spaces. This is because the action of a number of pollutants is to increase the dimensions of this space, in extreme cases leading to lifting-off of the epithelium. Estimation of volumes is carried out morphometrically by counting the points of a morphometric grid which fall on a particular tissue. Thus by counting the number of points falling on the lymphoid spaces and those on the whole secondary lamella an estimate is given of the proportion of the lamella occupied by these spaces. Similar procedures can establish for example, the proportionate volume of the blood spaces or the haematocrit value by counting points which fall on red blood cells relative to those which fall on the blood spaces (Hughes, 1979).

References

Hughes, G.M. (1979) The paths of blood flow through the gills of fishes - some morphometric observations. Bulgarian Academy of Sciences, Sofia, *Acta Morphologica.* 2:52-58.

Hughes, G.M. (1984) Measurement of gill area in fishes: Practices and problems. *J. Mar. Biol. Assoc.,* U.K. 64:637-655.

Hughes, G.M. (1989) Morphometry of respiratory systems. In: *Techniques in Comparative Respiratory Physiology: An Experimental Approach.* S.E.B. Seminar Series. 37:21-48. CU Press, Cambridge 1989.

Hughes, G.M., S.F. Perry (1976) Morphometric study of trout gills: A light microscopic method suitable for the evaluation of pollutant action. *J. Exp. Biol.* 64:447-460.

Hughes, G.M., Roy, P.K., Munshi, J.S.D. (1992) Morphometric estimation of oxygen diffusing capacity for the air-sac in the catfish. *Heteropneustes Fossilis J. Zoology.* 227:193-209.

Weibel, E.R. (1979) *Stereological Methods* Vol 1. Academic Press, New York.

Recording of Heart and Ventilatory Rhythms in Fishes

by G.M. Hughes
Research Unit for Comparative Animal Respiration, University of Bristol, Bristol BS8 1UG, U.K.

Principle

Electrophysiological techniques can provide useful information about the activity of certain organ systems, particularly those in which electrical activities are generated in excitable cells. The nervous and muscular systems are notable examples, e.g. the electroencephalogram of the brain; the electromyogram (EMG) of skeletal muscle makes it possible to establish the timing and extent of activity of particular muscles. The latter method has been used in fish to assess the degree of activity of the muscles which produce the breathing movements (Hughes and Ballintijn, 1965, 1968). The combined activity of these muscles results in pressure changes within the respiratory cavities and these are more conveniently recorded using pressure transducers which convert the pressure change to an electrical signal which can be recorded (Hughes and Shelton, 1958; Hughes, 1975). Respiratory movements can also be detected with various types of mechano-electrical transducers. Many of these are only usable if the fish is in a fixed position but there are others which, like the pressure recording technique, are possible in unrestrained animals. One particularly useful technique uses changes in impedance across the respiratory apparatus during these movements (Belaud *et al.*, 1971; 1977). Two electrodes are implanted one on either side of the opercular cavities and as these move the change in impedance between them is recorded by suitable electrical circuitry.

The electrocardiogram is well known in mammalian physiology where much information can be gained regarding heart function by studying the types of electrical signal which are generated. Apart from simple measurements such as frequency, the particular form of the electrical signal can give information about the coordination between different chambers of the heart (Satchell, 1971). The vertebrate heart is largely composed of cardiac muscle and the electrical activity spreads throughout the heart partly as a result of specially modified pathways which are more rapidly conducting, and ensure a greater synchrony of action in different parts of the ventricle, for example, which ensures a more effective pumping action. ECG's have been recorded from many animals including invertebrates, in which the coordination of the heart beat and its regulation is largely controlled by a localized cardiac ganglion. In the vertebrates the basic rhythm of the heart is governed by a myogenic pacemaker but this may be modified by the action of the autonomic system and by circulating chemical substances.

Procedures

Recording of the ECG

The electrical signals generated by the vertebrate heart are quite large especially as they usually involve more or less synchronous action of many parts of this organ. It is usually relatively easy to pick up such signals using modern amplifiers but it is important that the electrodes should be located in as good a position as possible. In this respect it is more or less easy to record the ECG in different species of fish. This demonstration uses two species, in one of which the heart is contained within a relatively thin pericardium whereas in the eel the pericardium is fairly thick, both from a mechanical and also an electrical point of view. In all cases it is necessary to localize the position of the heart in the particular fish to be used as the best signals for recording are obtained when the active tips of the electrodes are placed close to the heart. But care must be taken not to produce any damage which would interfere with its normal activity.

Preparation of electrodes

Bi-polar electrodes are usually used, that is two electrodes both connected to the input of the amplifier; the surrounding water is connected to earth (see Fig. 2). Under these conditions any signal producing a difference in electrical potential between the two electrode tips is amplified whereas identical changes at the two electrodes are cancelled out (common mode rejection). This latter is particularly important as it reduces interference due to pickup of the mains a.c. in the neighbourhood of the experimental set-up. As the potentials to be recorded are usually millivolts this is not as serious as in some experiments which involve recording of much smaller potential changes.

Different types of electrodes have been used for recording the electrocardiogram:

a) Small steel pins insulated at their tip. The small pins are soldered to insulated wire and the junction between the wire and most of the pin is covered in some epoxy cement such as 'araldite'. This is often done by dipping the pin into a dilute solution of araldite in chloroform and hanging it up to dry. If the pin has been properly cleaned and sharpened then the cement does not adhere to the fine tip and this forms the active lead for detecting potential changes. After drying the procedure may be repeated to thicken the insulation. Such electrodes have been used for EMG and ECG recording.

b) Fine stainless steel wire insulated with cement (diamel). This provides a very convenient material for electrodes implanted in small muscles or close to the heart of fishes. To do this each length of diamel-coated wire is taken and the insulation removed from the two ends by a very brief insertion in a flame. The tips are then cleaned using emery cloth or some other abrasive. At one end the length removed is about 1 cm and this is for connecting up to the amplifier leads. At the other end the wire is bared for only 1 or 2 mm. This is inserted into the anaesthetised fish for recording. To do the latter the wire is threaded through a fine hypodermic needle (gauge 25, no 18) and the tip is bent over (Fig. 1) and can then be inserted into the recording position and when the needle is withdrawn the hook of the electrode keeps it in position. The precise position and depth to which the needle is inserted depends on the particular fish.

Connection of the end of the recording wire to the amplifier is an important part of the arrangement and this may be achieved by using fine springs which are connected directly to the amplifier input leads; the stainless steel wire being clipped between 2 coils of the spring. Another method is to use a screw-type connector. Such methods are necessary because soldering of stainless steel wire is extremely difficult. The amplifier should have a high input impedance and be equipped with suitable filters. The ECG recordings are relatively low in frequency so that high frequency noise should be filtered out and frequencies up to 50 or 60 cycles should also be filtered in order to reduce any movement artefacts or mains hum.

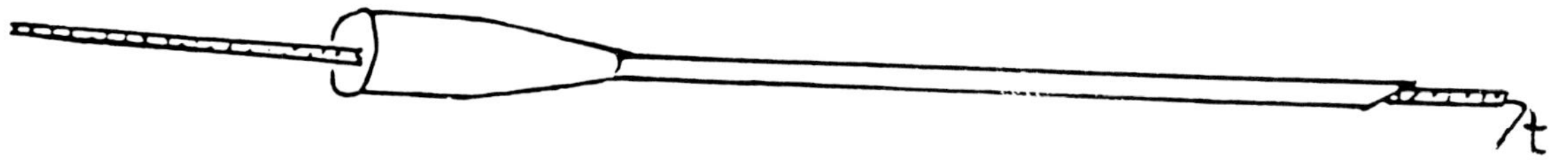

Fig.1. Diagram of an electrode made from diamel-coated stainless steel wire which has been inserted through a hypodermic needle and the insulating coating removed from the tip (t).

Recording of ECG in eels

As mentioned above, the eel pericardium is relatively thick so that recordings made in the manner described above are not so easily achieved as in fish such as rainbow trout and members of the carp family. The usual method that has been adopted by many authors is as follows. The anaesthetized (3% urethane) eel, possibly spinalized, is used for making a mid-ventral incision in the skin just in front of the pectoral fins. The skin and muscle are carefully withdrawn and the thick connective tissue of the pericardium exposed. This is further freed from the body wall musculature so that a wire can be threaded through from the inside of the incision to the outside using a suture needle threaded with strands of normal insulated wire. The internal end of this wire has been soldered to a blob of solder which is fitted into position close to the sides of the pericardium after the wire has been threaded through. Apart from the blob of solder there should be no other exposed portion of this lead. The mid-ventral incision is then stitched up and the fish allowed to recover from the anaesthetic.

The two leads are usually taken above the fish and tied into a loose knot and one or two stitches made to anchor the leads to the dorsal fin. The wires in this case are relatively thick and present no problem in connecting to the amplifier leads. It is possible to wind the leads round a cork which floats at the surface of the water and to which the shielded leads from the amplifier are led, being held above by a suitable clamp. Fig. 2 shows the way in which different pieces of equipment should be joined together. Notice that the water containing the fish must be earthed by a connection to the earth of the input lead of the preamplifier. The equipment should be connected to a true earth at one point. These are commonly observed routines to reduce interference with alternating current.

Ventilatory rhythm

When making recordings of ECG as described above it will have been noticed that sometimes the recording is distorted because of low frequency rhythmic movements of the base line. These can be filtered out if a recording of a pure ECG is required. The low frequency fluctuations are mainly artefacts produced by the ventilatory movements and can be used as a simple method of recording the rhythm but care must be taken to ascertain the number of fluctuations which correspond to a single ventilatory cycle. Just as the low frequency fluctuations may be filtered out for ECG recordings it is also possible to filter out the ECG and other artefacts and so produce a fairly clear recording of ventilatory frequency. Such sophisticated filtration circuitry has been employed in monitoring devices (Evans and Walwork, 1988) in which there is no attachment of electrodes to the fish which moves in a channel the sides of which form the electrode system.

As mentioned in the introduction the recording of ventilatory rhythm in the laboratory is best achieved by means of a catheter which records pressure changes within the buccal cavity. Such a catheter is conveniently made from polyethylene tubing which has been flared at the tip by placing its end close to a flame. A small piece of rubber with a central hole is then threaded down the catheter to the flared region. This small piece of rubber finally rests on the inside of the roof of the buccal cavity and causes less irritation to the fish. The catheter is inserted into a fully anaesthetized fish (MS 222 1g per 10 litres) using a hypodermic needle which is thrust through the roof of the mouth just behind the maxillary valve. The catheter is threaded through this needle and then held in position by another rubber pad and a short length of rubber tubing which keeps the pad in position on the outside of the fish (Fig. 3, r) The catheter should be as short as possible if accurate pressure recordings are to be made. Where the fish is contained in a small respirometer box this presents little difficulty. The end of the catheter is connected to a suitable pressure-electric transducer. The whole system between the buccal cavity and transducer sensor must be devoid of air bubbles if good recordings are to be made.

With this type of arrangement it is very easy to make separate recordings of the ventilatory rhythm, including coughing, of a fish over periods of several days during which different pollutants can be tested by inclusion in the water flowing into the respirometer system.

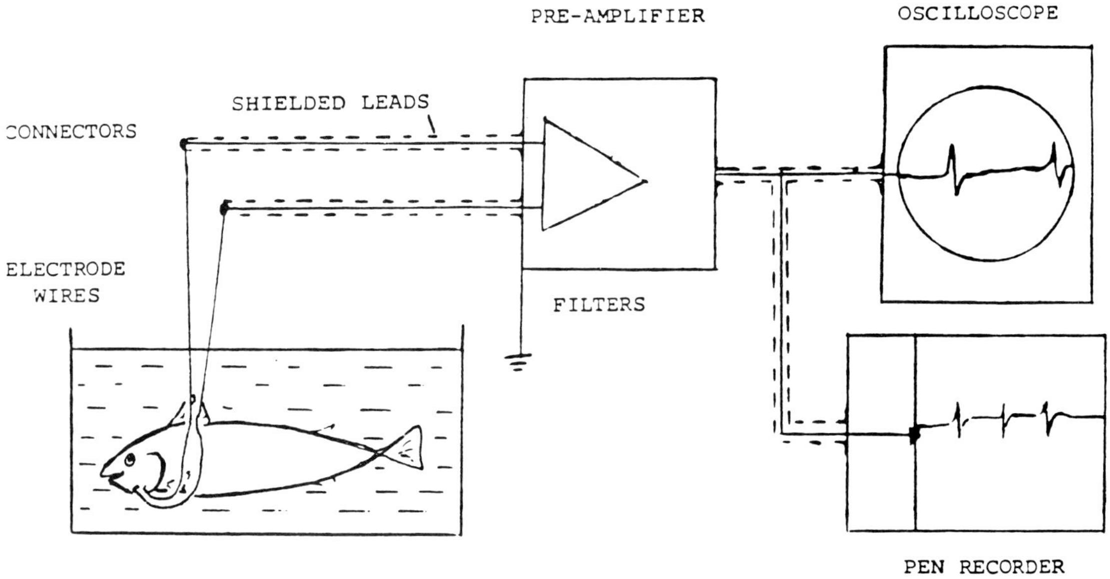

Fig.2. Diagram to show connections between electrodes and balanced inputs to a preamplifier provided with suitable high and low frequency filters. The output of the preamplifier is connected to the driver amplifier of an oscilloscope and/or pen recorder. If pressure recordings are made the preamplifer must be the appropriate one for the pressure transducer and its output can be fed to a second channel of the oscilloscope and/or pen recorder.

a)

b)

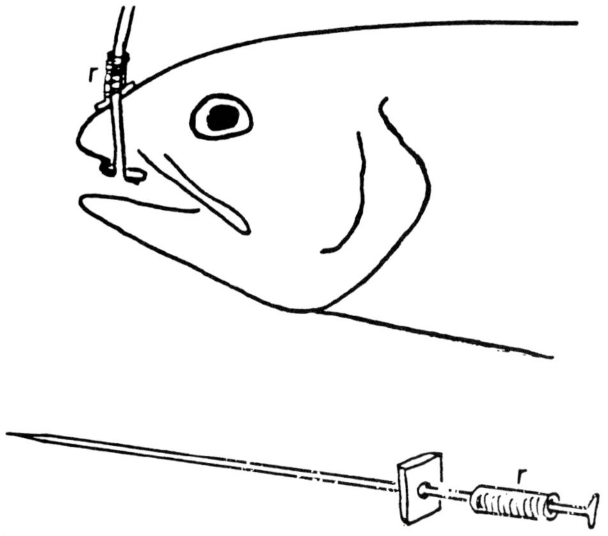

Fig.3. Diagrams to illustrate method of cannulating buccal cavity with a polyethylene catheter: (a) flared catheter has been inserted through the mouth and is held in position on the outside of the fish by a small rubber pad and a piece of suitable tubing (r); (b) shows a convenient way of keeping the rubber pad and tubing on a needle which can be inserted into the end of the catheter to facilitate the placement of the pad and tubing on to the polyethylene tubing.

Observations

When recordings of the ECG are being made it will be observed that they respond quite readily to disturbances of the fish. In some cases this can lead to a slowing of the heart (bradycardia) which in some species occurs even when the observer simply leans over the tank containing the fish. In other cases with increased activity of the fish the heart rate is usually higher.

It is clear, therefore, that before this type of preparation can be used for the assay of pesticides or other environmental factors the fish should be allowed to recover from the operation and to settle into the experimental arrangement. In order to reduce disturbances from outside the fish is often contained within a darkened chamber and there should be as little movement in the vicinity as possible. Only in this way can the changes in heart and ventilatory frequencies following the administration of any pollutant be taken as indicative of the effects of that agent. In practice the fish is often allowed to recover from the anaesthetic and acclimate in the experimental set-up for about 24 hours before any experimental changes are contemplated.

References

Belaud, A., Peyraud-Waitzenegger, M., Peyraud, C. (1971) Techniques originales de prelevement des parametres physiologiques sur le congre (*Conger conger* L) libre de ses deplacements. I. Electrocardiogramme et activite respiratoire. *Hydrobiologia.* 38:49-59.

Belaud, A., Sebert, P., Soulier, P., Peyraud, C. (1977) Relative estimation of ventilatory flow rate in the eel (*Anquilla anquilla* L). *Rev. Can. Biol.* 36:93-103.

Evans, G.P., Walwork, J.F. (1988) The W.R.C. fish monitor and other biomonitoring methods. In:

Automated Biomonitoring: Living Sensors as Environmental Monitors. Gruber, D.S., Diamond, J.M. (eds) pp. 75-90. Ellis Horwood, Chichester.

Hughes, G.M. (1975) Coughing in the rainbow trout, *Salmo gairdneri,* and the influence of pollutants. *Rev. Suisse Zool.* 82:47-64.

Hughes, G.M. (1985) Problems in assessing the effects of pollutants on the cardiorespiratory physiology of fishes. *J. Gen. Biol.* USSR Moscow 46:241-254 (Russian).

Hughes, G.M., Ballintijn, C.M. (1965) The muscular basis of the respiratory pumps in the dogfish, *Scyliorhinus canicula. J. Exp. Biol.* 43:363-383.

Hughes, G.M., Ballintijn, C.M. (1968) Electromyography of the respiratory muscles and gill water flow in the dragonet. *J. Exp. Biol.* 49:583-602.

Hughes, G.M., Shelton, G. (1958) The mechanism of gill ventilation in three freshwater teleosts. *J. Exp. Biol.* 35:807-823.

Satchell, G.H. (1971) *Circulation in Fishes.* pp. 1-131. Cambridge University Press.

Tort, L., Flos, R., Hughes, G.M. (1986) Metodos para el estudio de los efectos de la contaminacion en la fisiologia cardiorrespiratoria de los peces. *Inf. Tecn. Inst. Inv. Pesq.* 131:1-32.

Alterations in Growth, Reproduction, and Energy Metabolism of Estuarine Crustaceans as Indicators of Pollutant Stress

by Charles L. McKenney, Jr
U.S. Environmental Protection Agency, Environmental Research Laboratory,
Gulf Breeze, Florida 32561, U.S.A.

An estuarine mysid (*Mysidopsis bahia*) has been identified as one of the most sensitive members of the estuarine community to pollutant stress (for a review see Nimmo and Hamaker, 1982). In the majority of life-cycle toxicity tests using this planktonic estuarine crustacean, sublethal reduction in reproductive potential is the most sensitive criterion for chronic biological effect. Recent information (McKenney, 1982; 1985; 1986; McKenney and Mathews, 1990; McKenney *et al.*, 1991) suggests that retarded juvenile growth rates and alterations in energy metabolism of exposed mysids precedes reductions in reproductive capacity. Results of these laboratory studies, confirmed recently in a field study (McKenney *et al.*, 1985), indicate that measurements of metabolic dysfunction in mysids exposed to microcontaminants may be used to predict altered production rates in these sensitive crustacean populations.

Procedures for Reproductive Measurements

Newly released juvenile mysids (<24 h old) are obtained from ovigerous females using a recirculating seawater system similar to that described by Reitsema and Neff (1980). Fifteen juveniles are pipetted into Nitex* baskets constructed of 15 cm glass Petri dishes to which a cylinder of nylon screen (mesh 210 µm) is attached by silicone cement. Each basket is placed in one of 3 replicate chambers for each exposure condition in a flow-through exposure system (see McKenney, 1986 for a more detailed description) using either a proportional diluter or a syringe pump to inject known concentrations of the toxicant (Schoor and McKenney, 1983). Similar groups of mysids are placed in additional baskets at each exposure concentration for subsequent subsampling for growth and metabolic measurements.

Throughout the study, all groups of mysids are fed freshly hatched *Artemia nauplii* (<24 hours old) at densities (2-3 *A. nauplii* per ml) previously determined to maximize growth and reproduction (McKenney, 1987).

Exposed mysids are observed daily for survival and dead animals are removed. As juveniles mature, an ovigerous female is paired with a mature male within a Nitex brood cup constructed of a 6 cm glass Petri dish to which a nylon screen cylinder is attached. Daily records are maintained for day of release of first brood and number of young released in the first brood of isolated females.

---

*Mention of commercial products does not imply endorsement by the U.S. Environmental Protection Agency.

This study is terminated 7 days after the mean day of release of the first brood by control mysids, since experience has shown that toxicant-induced delay in release of first brood occurs within 7 days. Reproductive responses include mean day of release of the first brood, number of young per female in the first brood, total number of young per replicate population, and percentage of reproductively active females in the population.

Procedures and Calculations for Measurements of Energy Metabolism

On days representative of the various mysid life stages (for *Mysidopsis bahia* reared at 25°C and 2 ^{0}/00 S: Day 0, newly released juvenile; Day 4, early juvenile; Day 9, juvenile; Day 16, advanced juvenile/young adult) (McKenney, 1986). Eight mysids from each exposure condition are sealed individually in 5 ml all-glass syringes containing 1 to 3 ml (dependent on the size of the mysid) of fully oxygenated media at the appropriate exposure concentration. Syringes containing mysids are incubated in a 25°C water bath with control syringes without mysids. After 4 to 6 hours, 75-100 µl of the fluid in the syringe is injected into a Radiometer BMS3 Mk2 Blood Micro System with attached PHM71 Mk2 Acid-Base Analyser and a direct measurement is recorded for partial pressure of oxygen. In addition, 100 µl of the fluid is injected into a micro-flow-through cap (Model 95-00-25) attached to an Orion Ammonia Electrode (Model 95-10) connected to an Orion Research Microprocessor Ionalyser (Model 901). A direct reading of ammonia-nitrogen concentration is recorded from the Ionalyser, which was calibrated in the range of experimental samples using ammonium chloride standards. Each mysid is then removed from the syringe, rinsed briefly in distilled water and placed in an oven at 60°C to dry for 48 hours. Dry weights of individual mysids are subsequently measured to the nearest 0.1 µg on a Cahn 21 Automatic Electrobalance.

Weight-specific respiration rates are determined using the equation

$$VO_2 = \frac{((1000 \times BC)/BP \times pO_2i) - ((1000 \times BC)/BP \times pO_2f) \times (V/1000)}{(DW/1000)/t}$$

where VO_2 = the respiration rate (µl O_2 /mg dry wt/h);
BC = Bunsen coefficient (µl O_2/ml/atm) for the solubility of oxygen at a particular salinity and temperature (Weiss, 1970); BP = barometric pressure (mm Hg); pO_2i and pO_2f = partial pressures (mm Hg) of oxygen in control syringes and in syringe containing mysid, respectively; V = volume (ml) of seawater in syringe; CW = dry weight (mg) of mysid; and t = time (h) in syringe. Multiplying the VO_2 value by 1.43 converts the weight-specific rate of oxygen consumption to µg O_2/mg dry wt/h.

Weight specific ammonia excretion rates (μg NH_3-N/mg dry wt/h) are calculated from the differences between ammonia-nitrogen concentrations in control syringes and syringes containing a mysid. O:N ratios are calculated as the ratio of atoms of oxygen consumed to atoms of nitrogen excreted. A theoretical O:N ratio of 7 closely approximates wholly protein catabolism by planktonic marine crustaceans (Mayzaud, 1973). A higher proportion of carbohydrates and/or lipids utilized as substrate by the organism increases the value of the O:N ratio.

Calculations for Growth Measurements

Several growth indices are calculated by the formulas of Winberg (1971) and Winberg *et al.* (1971) using dry weights of the various mysid life stages under each exposure condition. The mean specific rate of growth is calculated by the equation

$$gn = \frac{(\ln W_{n+1} - \ln W_n)}{t}$$

where g_n = the mean specific rate of growth of a particular life stage n; W_n and W_{n+1}= the mean dry weight of life stages n and n+1, respectively; and t = the number of days between life stage n and life stage n+1. The mean specific rate of growth is incorporated into the formula

$$G_n = (e^{g_n} - 1)$$

to quantify the fractional daily weight gain (G_n) for each life stage under the various exposure conditions. G_n is multiplied by the replicate dry weights to produce values of daily weight gain (WG) (μg/d) for each life stage-exposure condition combination.

Calculations for Net Growth Efficiency

Daily production energy for each life stage under the various exposure conditions is determined by the formula

$$P_n = (WG_n \times TC) + 0.10\ (WG_n \times TC)$$

where P_n = gross daily production in joules for life stage n, WG_n = daily weight gain (μg/d) for life stage n, TC = the mean energy content for dry tissue (18.598×10^{-3} joules per μg dry weight for an estuarine mysid from Florida, USA) (Johnson and Hopkins, 1978) and 0.10 (WG_n x TC) = a correction factor for the average loss from marine crustaceans by shedding of exuivae (McKenney, 1982 and references therein).

Daily maintenance energy for each life stage is derived from the formula

$$R_n = VO_{2n} \times 24 \times W_n \times OC$$

where R_n = daily expenditure in joules for maintenance during life stage n, VO_{2n} = the oxygen consumption rate (µg O_2/mg dry wt/h) for life stage n, W_n = dry weight (mg) of life stage n and OC = the oxycalorific coefficient in joules for carnivores that utilize ammonia as their chief excretory product (13.556×10^{-3} joules per µg O_2) (Elliott and Davison, 1975). Net growth efficiency (K_2) (Ivlev, 1945; Winberg, 1971; Winberg *et al.*, 1971) is derived for each life stage under each exposure condition by the formula

$$K_{2n} = \frac{P_n}{(P_n - R_n)}$$

where K_{2n} = the net growth efficiency for life stage n, P_n = the daily production in joules for life stage n, and R_n = the daily expenditure for maintenance in joules for life stage n. Net growth efficiency values represent the percentage of assimilated energy converted into growth of new tissue. A reduction in K_2 would indicate a stress condition for the organism with greater metabolic demands for maintenance leaving less energy for tissue production.

References

Elliott, J.M., Davison, W. (1975) Energy equivalents of oxygen consumption in animal energetics. *Oecologia.* 19:195-201.

Ivlev, V. S. (1945) The biological productivity of waters. *Usp. sourem. Biol.* 19 (Translated by W. E. Ricker. 1966. *J. Fish. Res. Board Can.* 23:1727-1759).

Johnson, J.T., Hopkins, T.L. (1978) Biochemical components of the mysid shrimp *Taphromysis bowmani* Bacescu. *J. Exp. Mar. Biol. Ecol.* 31:1-9.

Mayzaud, P. (1973) Respiration and nitrogen excretion of zooplankton. II. Studies of metabolic characteristics of starved animals. *Mar. Biol.* 21:19-28.

McKenney, C.L., Jr (1982) Interrelationships between energy metabolism, growth dynamics, and reproduction during the life cycle of *Mysidopsis bahia* as influenced by sublethal endrin exposure. In: *Physiological Mechanisms of Marine Pollutant Toxicity.* Vernberg, W.B., Calabrese, A., Thurberg, F.P., Vernberg, F.J. (eds). Academic Press, New York, pp. 447-476.

McKenney, C.L., Jr (1985) Associations between physiological alterations and population changes in an estuarine mysid during chronic exposure to a pesticide. In: *Marine Pollution and Physiology: Recent Advances.* Vernberg, F.J., Thurberg, F.P., Calabrese, A., Vernberg, W.B. (eds). University of South Carolina Press, Columbia, pp. 397-418.

McKenney, C.L., Jr (1986) Influence of the organophosphate insecticide fenthion on *Mysidopsis bahia* exposed during a complete life cycle. I. Survival, reproduction, and age-specific growth. *Dis. Aquat. Org.* 1:131-139.

McKenney, C.L., Jr (1987) Optimization of environmental factors during the life cycle of *Mysidopsis bahia*. Environmental Research Brief, EPA/600/M-87/004, U.S. EPA, Cincinnati, OH, 6 pp.

McKenney, C.L., Jr, Matthews, E., Lawrence, D.A., Shirley, M.A. (1985) Effects of ground ULV application of fenthion on estuarine biota. IV. Lethal and sublethal responses of and estuarine mysid. *J. Fl. Anti-Mosq. Assoc.* 56:72-75.

McKenney, C.L., Jr, Mathews, E. (1990) Alterations in the energy metabolism of an estuarine mysid (*Mysidopsis bahia*) as indicators of stress from chronic pesticide exposure. *Mar. Environ. Res.* 30: 1-19.

McKenney, C.L., Hamaker, T.L., Mathews, E. (1991) Changes in the physiological performance and energy metabolism of an estuarine mysid (*Mysidopsis bahia*) exposed in the laboratory through a complete life cycle to the defoliant DEF. *Aquat. Toxicol.* 19:123-135.

Nimmo, D.R., Hamaker, T.L. (1982) Mysids in toxicity testing - a review. *Hydrobiologia.* 93:171-178.

Reitsema, L.A., Neff, J.M. (1980) A recirculating artificial seawater system for the laboratory culture of *Mysidopsis almyra* (Crustacea: Pericaridea). *Estuaries* 3:321-323.

Schoor, W.P., McKenney, C.L., Jr (1983) Determination of fenvalerate in flowing-seawater exposure studies. *Bull. Environ. Contam. Toxicol.* 27:256-267.

Weiss, R.F. (1970) The solubility of nitrogen, oxygen and argon in water and seawater. *Deep Sea Res.* 17: 721-735.

Winberg, G.G. (1971) *Methods for the Estimation of Production of Aquatic Animals.* Academic Press, New York. 175 pp.

Winberg, G.G., Patalas, K., Wright, J.C., Hillbricht-Ilkowska, A., Cooper, W.E., Mann, K. H. (1971) Methods for calculating productivity. In: *Secondary Productivity in Fresh Waters.* Edmondson, W.T., Winberg, G.G. (eds). Blackwell Scientific Publications, Oxford. pp. 296-317.

A Fish Model as an Indicator for Teratogenic Substances

by Douglas P. Middaugh and Michael J. Hemmer
U.S. Environmental Protection Agency, Environmental Research Laboratory,
Sabine Island, Gulf Breeze, Florida 32561, U.S.A.

A recent expansion has occurred in development of *in vivo* methods for teratogenesis (Birge *et al.*, 1983; Whitby and Flynn, 1987). This effort has focused on a number of potential nonmammalian systems, and fish have received considerable attention (Laale and Lerner, 1981; Collins, 1987; Weis and Weis, 1987; Solomon and Faustman, 1987). We have developed a fish embryo model as an indicator for teratogenic substances. Our procedure utilizes naturally spawned embryonic inland silversides, *Menidia beryllina.* This euryhaline fish is ubiquitous in estuaries, coastal rivers and numerous lakes in the southeastern United States and is also found in the Mississippi River basin (Sisk and Stephens, 1964; Gomez and Lindsay, 1972; Johnson, 1975; Chernoff *et al.*, 1981).

Embryo Acquisition

Adult *M. beryllina* are easily maintained in the laboratory. A population of 25 males and 25 females collected from a low salinity (0-5 ^{0}/oo) estuarine site in early April (20°C) and maintained at 5^0/oo 25°C produced 300 to 2000 embryos daily for approximately 5 months (Middaugh *et al.,* 1985). Thereafter a new population of young adults collected in late September, produced embryos throughout the winter. Holding tank configuration, environmental and nutritional requirements for brood stock adults have been summarized by Middaugh *et al.* (1986).

Test Procedure

Natural unpolluted seawater used in all tests is treated by pre-filtration through a 20 µm polypropylene filter. It is then passed through a 6 µm filter, diluted to 5^0/oo salinity with deionized water and autoclaved. Stock solutions are prepared by diluting the analytical grade test chemical in deionized water, acetone, or another appropriate solvent.

A range finder test may be employed using a reduced number of embryos and a wide range of concentrations to bracket the potential teratogenic response levels. It is preferable that a concentration range be chosen that limits the lethal response at the highest treatment levels while enhancing the teratogenic response at the lower treatment levels.

Just prior to beginning a test, aliquots of stock solution are added to 1.0 litre of autoclaved saline water (5^0/oo) to yield the desired nominal exposure concentrations. We generally use 5 concentrations of the suspected teratogen plus a seawater control. If an organic solvent such as acetone is employed, an additional carrier control must be maintained with the concentration of the solvent equal to that in the highest exposure concentration of teratogen. The quantity of solvent added to each treatment should be equal to that in the highest exposure concentration of toxicant and in the carrier-control; and should not exceed concentrations recommended by ASTM (1980).

Exposures are conducted in borosilicate glass tissue culture tubes 16 x 93 mm (window size 11 x 55 mm). Twenty tubes per concentration are used, including 20 controls and 20 carrier controls (if required). Numbered tubes are randomized prior to addition of embryos.

Early blastula embryos, rinsed 3 times with autoclaved $5^0/oo$ water, are placed in a crystallization dish under a dissecting scope (15-20x). Single embryos are drawn into a pasteur pipette and placed in an empty tube with ~ 0.1 ml of $5^0/oo$ water from the crystallization dish. The randomized numbered tubes, each containing a single embryo, are then quickly reordered (1-20, 21-40, 41-60, 61-80, 81-100, 101-120, 121-140) and 6 ml of respective exposure solutions (or clean $5^0/oo$ saline water for controls) are added to each tube. Tubes, each with an airspace of 7 cc, are sealed with teflon-lined caps, placed in racks and stored in a horizontal position in an incubator at 25± 1°C with a 14L:10 D photoperiod. Cool-white fluorescent lamps provide 500 lux illumination during the light phase.

Dissolved oxygen (D.O.) and pH should be checked in 3 to 5 control(s) and respective treatment tubes at the end of each test. A Lazar Model DO 166* oxygen probe and Orion Model SA 520 pH meter, or equivalent, are suitable for measurements. In conducting ~ 50 tests with a variety of teratogens, including aliphatic and aromatic hydrocarbons, measured pH values have ranged from 6.1 to 7.3 and D.O. from 5.1 to 7.6 mg/l. Maintenance of adequate D.O. concentrations is essential since low D.O. can also cause terata.

Quantification of Responses

Individual control and chemically-exposed embryos are examined daily for 7 days until death or hatching occurs. The criterion for a successful test is 85% or greater hatching in the control and carrier control. A Zeiss Invertoscope D Microscope, or equivalent, is used to categorize responses. Observed terata are scored each day using a numerical severity index for craniofacial (CR), cardiovascular (CV), and skeletal (SK) defects. Numerical scores (Table 1) are based upon classification schemes devised by Weis and Weis (1977), Weis *et al.* (1981), and Weis and Weis (1982).

Data Analysis

Upon completion of a test embryo deaths are categorized based on percentage mortality and analysed, by either probit analysis to derive an LC50 or by a one-way analysis of variance (ANOVA) and Duncans multiple range test, using arcsine transformed data to

---

*Mention of tradenames does not imply endorsement by the U.S. Environmental Protection Agency.

indicate differences between control(s) and treatment responses (Finney, 1978; Steel and Torrie, 1960; Sokal and Rohlf, 1969). It is preferable that a concentration range be chosen that limits the lethal response at the highest treatment level while enhancing the teratogenic response.

Analysis of the total teratogenic response is performed as follows: for each tube, CR, CV, and SK values (Table 1) are summed to provide a total daily score which can range from 0 for a normally developing embryo to 13 for a severely deformed embryo (i.e., CR 3 + CV 5 + SK 5 = 13). If an embryo dies before the end of the test, a value of 14 is assigned to the tube daily from the date of death until test termination. The daily scores are then summed to generate a total severity index for each embryo. By assigning a value of 14 to exposed and control embryos that die, the uncertainty of causal effect for death is accounted for without biasing the total severity-indices for terata (Middaugh *et al.*, 1988). To test for differences in terata between control(s) and treatment concentrations, nonparametric Van der Waerden normal scores tests (SAS, 1985) and Van der Waerden post-hoc multiple comparison procedures, $\alpha = 0.05$ (Marascuilo and McSweeney, 1977) are conducted, using total severity-indices.

Analysis of individual response categories, CR, CV and SK (Table 1), may be conducted as follows: for each embryo, the daily score for the CR index could range from 0 to 3; for CV, 0 to 5; and for SK, 0 to 5. If an embryo dies before the end of the test, a daily value of 4 is assigned to the CR index; the CV and SK indices are assigned a value of 6. As with total severity-indices, assignment of a daily value for embryo death in individual categories enables one to account for mortalities without introduction of bias by mortalities of unknown cause (Middaugh *et al.*, 1986). Respective categorical indices are then analysed using Van der Waerden normal scores tests (SAS, 1985) and post-hoc multiple comparison procedures (Marascuilo and McSweeney, 1977).

With some compounds, the teratogenic expression is limited to one dysmorphic response, e.g., skeletal defect. In this case, the data analysis is greatly simplified. The teratogenic response is analysed by calculating the percentage of embryos that elicit a response at each treatment level. The data can then be presented graphically or arcsine transformed and analysed by one-way ANOVA and post-hoc procedures, if appropriate. If categories based on the severity of the response can be distinguished with the single morphogenic response, then the analysis is the same as described above for individual response categories (CR, CV and SK).

Utilization of the system described above provides for quantification of teratogenic/toxic effects of chemicals during embryonic development. Because the exposure tubes are sealed for the duration of the test, the risk of human exposure to the test compounds is greatly reduced. Furthermore, embryos from other species such as the Japanese medaka, *Oryzias latipes*, and zebra fish, *Brachydanio rerio*, probably could be tested in this system.

Table 1. ***Menidia beryllina.*** **Synopsis of observed teratological responses in embryos and numerical severity-index for craniofacial (CR), cardiovascular (CV), and skeletal (SK) defects. Adapted in part from Weis and Weis (1977), Weis** ***et al.*** **(1981), and Weis and Weis (1982).**

Craniofacial		Cardiovascular		Skeletal	
Value	**Effect**	**Value**	**Effect**	**Value**	**Effect**
0	none observed	0	none observed	0	none observed
1	slight defect in structure or size	1	slight defect in structure or function including reduced circulation	1	slight bend or kink including scoliosis, lordosis
2	moderate defect in structure or size including synophthalmia	2	tube heart, beating with or without circulation	2	major bend or kink (greater than 90° angle or more than one bend)
3	severe defect in structure or size including microphthalmia, anophthalmia, or anencephaly	3	tube heart, not beating	3	stunted
		4	beating tissue, but no heart structure	4	very stunted, but axis discernible
		5	no discernible heart	5	no axis discernible

References

American Society for Testing and Materials. (1980) Standard practices for conducting acute toxicity tests with fishes, macroinvertebrates and amphibians. In: *Annual Book of ASTM Standards.* pp. 1-25. ASTM, Philadelphia.

Birge, W.J., Black, J.A., Westerman, A.G., Ramey, B.A. (1983) Fish and amphibian embryos - a model system for evaluating teratogenicity. *Fundamentals Appl. Toxicol.* 3:237-242.

Chernoff, B., Conner, J.V., Bryan, C.F. (1981) Systematics of the *Menidia beryllina* complex (Pisces: Atherinidae) from the Gulf of Mexico and its tributaries. *Copeia.* 1981:319-335.

Collins, T.F.X. (1987) Teratological research using *in vitro* systems. V. Nonmammalian model systems. *Environ. Health Perspectives.* 72:237-249.

Finney, D.J. (1978) *Statistical Methods in Biological Assay*, 3rd Edition. Charles Griffin and Co. Ltd., London. 508 pp.

Gomez, R., Lindsay, H.L,. Jr (1972) Occurrence of the Mississippi silverside, *Menidia audens* (Hay) in Keystone Reservoir and the Arkansas River. *Proc. Okla. Acad. Sci.* 52:16-18.

Johnson, M.S. (1975) Biochemical systematics of the Atherinid genus *Menidia. Copeia* 1975: 662-691.

Laale, H.W., Lerner, W. (1981) Teratology and early fish development. *Am. Zool.* 21:517-533.

Marascuilo, L.A., McSweeney, M. (1977) *Nonparametric and Distribution - Free Methods for the Social Sciences.* Brooks/Cole, Monterey, CA. 556 p.

Middaugh, D.P., Hester, P.G., Meish, M.V., Stark, P.M. (1985) Preliminary data on use of the inland silverside, *Menidia beryllina,* to control mosquito larvae. *J. Am. Mosquito Control Assoc.* 1:435-441.

Middaugh, D.P., Hemmer, M.J., Rose, Y.L. (1986) Laboratory spawning of the inland silverside, *Menidia beryllina,* and Tidewater Silverside, *Menidia peninsulae,* with notes on survival and growth of larvae at different salinities. *Environ. Biol. Fishes.* 15:107-117.

Middaugh, D.P., Hemmer, M.J., Lores, E.M. (1988) Teratological effects of 2,4-dinitrophenol, "produced water" and naphthalene on embryos of the inland silverside, *Menidia beryllina. Dis. aquat. Org.* 4. 53-65.

SAS Institute Inc. (1985) *SAS User's Guide: Statistics,* Version 5 Edition. Carey, NC. 956 p.

Sisk, M.E., Stephens, R.E. (1964) *Menidia audens* (Pisces: Atherinidae) in Boomer Lake, Oklahoma, and its possible spread in the Arkansas River System. *Proc. Okla. Acad. of Sci.* 44:71-73.

Sokal, R.R., Rohlf, F.J. (1969) *Biometry. The Principles and Practice of Statistics in Biological Research.* W.H. Freeman Co., San Francisco. 776 pp.

Solomon, F.P., Faustman, E.M. (1987) Developmental toxicity of four model alkylating agents on Japanese medaka fish (*Oryzias latipes*) embryos. *Environ. Toxicol. Chem.* 6:747-753.

Steel, R.G.D., Torrie, J.H. (1960) *Principles and Procedures in Statistics.* McGraw-Hill, New York. 481 p.

Weis, J. S., Weis, P. (1977) Effects of heavy metals on development of the killifish, *Fundulus heteroclitus, J. Fish. Biol.* 11:49-54.

Weis, J.S., Weis, P., Ricci, J.L. (1981) Effects of cadmium, zinc, salinity and temperature on the teratogenicity of methylmercury to the killifish, *Fundulus heteroclitus. Rapp. P.-V. Reun. Cons. Int. Explor. Mer.* 178:64-70.

Weis, P., Weis, J.S. (1982) Toxicity of methylmercury, mercuric chloride and lead in killifish, *Fundulus heteroclitus,* from Southampton, New York. *Environ. Res.* 28:364-374.

Weis, J.S., Weis, P. (1987) Pollutants as developmental toxicants in aquatic organisms. *Environ. Health Perspect.* 71:77-85.

Whitby, K.E., Flynn, T.J. (1987) Introduction - teratological research using *in vitro* systems. *Environ. Health Perspect.* 72:201

Enzymatic Determinations as Tools for the Detection of Water Pollution in Fishes

by J. Nemcsøk
Department of Biochemistry, Jozsef Attila University, Szeged, Hungary

Principle

Water pollution caused by chemical compounds occurs ever more frequently in the world. The main sources of pollution are agriculture, petroleum refining, domestic sewage and chemical manufacturing. The adverse effect of such pollution on aquatic life falls into three categories: the direct toxicity of the chemicals to aquatic organisms, the lowering of dissolved oxygen in receiving waters, and the introduction of bad taste and odour to edible flesh of fish (Phipps *et al.*, 1981; Olàh *et al.*, 1986). Because of the wide, high volume usage of pesticides in agriculture, it is to our benefit to be aware of its toxicity in the aquatic environment. This awareness must include a knowledge of the effects on fishes, more specifically on some of their biochemical parameters.

The pollution of natural water may cause significant tissue damage in fish (McKim *et al.*, 1970; Reichenbach-Klinke, 1972; Horvàth and Stammer, 1979; Ferri and Macha, 1980; Rojik *et al.*, 1983; Benedeczky *et al.*, 1984). Enzymes getting into the blood after cell necrosis of certain organs may be used to indicate the tissue damage (Kristofferson *et al.*, 1974; Nemcsøk *et al.*, 1981; Nemcsøk and Boross, 1982).

LDH (lactate dehydrogenase) has been used for demonstrating tissue damage in fish for a long time (Kristofferson *et al.*, 1974; Nemcsøk and Boross, 1982). The increased LDH level in blood sera, however, shows only the degree of tissue damage without letting us know which tissue or organ necrosis is responsible for the increased LDH activity levels of blood sera.

For screening the effects of organophosphate and carbamate insecticides on fish acetylcholinesterase (AChE) inhibition can be regarded as good indicator of exposure (Holland *et al.*, 1967). Most of these insecticides inhibit cholinesterase activity at low concentration (Coppage, 1971; Coppage and Matthews, 1975; Coppage *et al.*, 1975; Coppage and Braidech, 1976; Klaverkamp *et al.*, 1977; Duangsawasdi and Klaverkamp, 1979; Klaverkamp and Hobden, 1980).

Transaminase Enzymes

Introduction

In human diagnosis the detection of injuries of liver, kidney and muscle tissues by determination of transaminase activity in blood is a commonly used, well proven method. Transaminases are located in cytoplasm and in mitochondria under normal conditions. The damages and lysis of the cells result in these enzymes entering the blood in relatively high quantities. This causes a rapid increase of enzyme activity "blood transaminase". Some laboratories have found that the same method can be used in detecting tissue injuries in some fish species, which have been caused by the pollution of their environment.

Method (Reitman and Frankel, 1957)

Sample material

Serum

Reagents

	Contents	Initial concentrations of solutions
1a	phosphate buffer	0.1 mol/l, pH 7.4
	L-aspartate	0.1 mol/l
	α-oxoglutarate	2.0 mmol/l
1b	phosphate buffer	0.1 mol/l, pH 7.4
	D, L-alanine	0.2 mol/l
	α-oxoglutarate	2.0 mmol/l
2	2,4-dinitrophenyl-hydrazine	1.0 mmol/l

Additional reagent

Sodium hydroxide, ca. 0.4 N.

Preparation and stability of solutions

1a, 1b and 2	Use solutions undiluted. Keep bottles tightly closed. Stable up to the expiry date specified when stored at + 15 to 25°C.

Sample preparation

Hemolysis interferes with the test.
Use only fresh serum.

Procedure (Table 1)

Wavelength: Hg 546 nm (530-550 nm)
Cuvette: 1 cm light path
Incubation temperature: 37°C
Measure against blank.

One blank is sufficient for each assay series.

Table 1.

Pipette into test tubes				
	blank		sample	
	GOT	GPT	GOT	GPT
solution 1a	0.5ml	-	0.5ml	-
solution 1b	-	0.5ml	-	0.5ml
serum	-		0.1 ml	
redist.water	0.1 ml		-	
Mix, incubate in water bath at 37°C for exactly 1 hour (GOT test) or exactly 30 min (GPT test). Add				
solution 2	0.5 ml		0.5 ml	
Mix, let stand at 20-25°C for exactly 20 min. Add				
sodium hydroxide solution	5.0 ml		5.0 ml	
Mix. Pour into cuvettes after 5 min, read absorbance of sample against (A).				

If the absorbance A exceeds: 0.260 (GOT assay), or 0.375 (GPT assay), dilute 0.1 ml serum with 0.9 ml of 0.62% NaCl solution and repeat assay with 0.2 ml of this dilution (result x 10).

Calculation

Obtain GOT and GPT activities in the serum from the table of values below.

Table of values for measurement at Hg 546 nm

GOT		GPT	
A	U/l	A	U/l
0.020	4	0.025	2
40	8	50	5
60	11	75	9
80	16	0.100	12
0.100	19	125	16
120	24	150	20
140	28	175	24
160	34	200	29
180	40	225	34
200	48	250	39
220	57	275	45
240	67	300	52
0.260	80	325	59
		350	68
		0.375	77

Acetylcholinesterase Enzyme

Introduction

The physiological significance of acetylcholinesterase (choline esterase) is due to the rapid cleavage of acetylcholine, which has a high biological activity, into choline and acetic acid. Acetylcholine is essential for the conductance of nerve impulse and the generation of bioelectrical current. The enzyme is found in the nerve tissue and erythrocytes of mammals, in snake venom and in the tissue and serum of some fishes.

Measurements on the changes in enzyme activity of acetylcholinesterase (AChE, EC 3.1.1.7.) have been used for a long time determining the characteristic toxic effect of insecticides containing organic phosphate and of carbamate type (Gage, 1955; Gage, 1967; Haet, 1961; O'Brien, 1960). It has also been demonstrated in the case of fish that water contamination may also cause damage in the nervous system through the inhibition of acetylcholinesterase (Holland *et al.*, 1967; Coppage *et al.*, 1975; Coppage and Braidech, 1976; Klaverkamp *et al.*, 1977; Klaverkamp and Hobden, 1980; Hanke *et al.*, 1983; Nemcsøk *et al.*, 1984).

Assay principle

$$\text{(1) acetylcholine} + H_2O \xrightarrow{\text{acetylcholinesterase}} \text{acetate} + \text{thiocholine}$$

$$\text{(2) thiocholine+dithiobisnitrobenzoate} \rightleftharpoons \text{thiocholine-thionitrobenzoate disulphide+thionitrobenzoate}$$

The change in optical density at 405 nm per unit time is a measure of the acetylcholinesterase activity.

Solutions

I. Phosphate buffer, 52 mM; pH=7.2 (193,9 μg $KH_2PO_4.H_2O$), 1312 mg Na_2HPO_4 + 10.28 mg 5.5' dithio-bis-2-nitrobenzoic acid (100 ml).

II. Acetylthiocholine iodide solution 82.4 mM, 0.156 M/45.1 mg acetylthiocholine iodide (ml)

Sample: 1mg enzyme lyophilized material per ml, diluted 1:10,000 with redist. water immediately before the measurement.

Spectrophotometric measurement:

Wavelength: 412 nm; light path: 1 cm; final volume: 3.12 ml;
Temperature: 25°C; read against blank.

Table 2.

Pipette into cuvette		Concentration in the assay mixture
phosphate buffer (I) acetylthiocholine iodide solution (II)	3.00 ml 0.10ml	approx. 48 mM 5 mM
mix, start reaction by addition of		
sample	0.02ml	
mix, read optical density after 1,2,3, 4 and 5 min. Calculate the mean from the measured ΔE/min. values.		

Calculation

$$\text{Volume activity} = \frac{3.12}{13.3 \times 1.0 \times 0.02} \; \Delta E/\text{min (U/ml sample)}$$

$$\text{Specific activity} = \frac{\text{volume activity}}{\text{concentration}} \; \text{(U/mg protein)}$$

References

Benedeczky, L., Birò, P., Schaff, Zs. (1984) Effect of 2,4-D containing herbicide Diconirt on ultrastructure of carp liver cells. *Acta Biologica Szegediensis.* 30:107-127.

Coppage, D.L. (1971) Characterization of fish brain acetylcholin-esterase with an automated pH stat for inhibition studies. *Bull. Environ. Contam. Toxicol.* 6:304-310.

Coppage, D.L., Matthews, E. (1975) Brain acetylcholinesterase inhibition in a marine teleost during lethal and sublethal exposures to 1,2-dibromo-ethyl phosphate (Naled) in seawater. *Toxicol. Appl. Pharmacol.* 31:128-133.

Coppage, D.L., Matthews, E., Cook, G.H., Knight, J. (1975) Brain acetylcholinesterase inhibition in fish as a diagnosis of environmental poisoning by malathion. *Pestic. Biochem. Physiol.* 5:536-542.

Coppage, D.L., Braidech, T.E. (1976) River pollution by acetylcholinesterase agents. *Water Research,* 10, 19

Duangsawasdi, M., Klaverkamp, J.F. (1979) Acephate and fenitrothion toxicity in rainbow trout: Effect of temperature stress and investigations on the sites of action. In: *Aquatic Technology,* ASTM STP 667, Marking, L.L., Kimerle, R.A. (eds) American Society for Testing and Materials, 35.

Ferri, S., Macha N. (1980) Lysosomal enhancement in hepatic cells of teleost fish induced by cadmium. *Cell Biology International Reports.* 4:357-363.

Gage, J.C. (1955) Blood cholinesterase values in early diagnosis of excessive exposure to phosphorus insecticides. *Brit. Med. J.* 1:1370.

Gage, J.C. (1967) The significance of blood cholinesterase activity measurements. *Residue Reviews.* 18:159-173.

Haet, D.F. (1961) *Organophosphorus Poisons, Anticholinesterase and Related Compounds.* Pergamon Press, New York.

Hanke, W.G., Gluth, G., Bubel, H., Müller, R. (1983) Physiological changes in carps induced by pollution. *Ecotoxicology and Environmental Safety.* 7:229-241.

Holland, H.T., Coppage, D.L., Butler, P.A. (1967) Use of fish brain acetylcholinesterase to monitor pollution by organo- phosphorus pesticides. *Bull. Environ. Contam. Toxicol.* 2:156-162.

Horvàth, I., Stammer, A. (1979) Electron microscopical structure of gill lamellas of the ide (*Leuciscus idus*) with particular regard to the chloride cells and H_2S pollution. *Acta Biologica Szegediensis.* 25:133-142.

Klaverkamp, J.F., Duangsawasdi, M., MacDonald, W.A., Majewski, H.S. (1977) An evaluation of fenitrothion toxicity in four life stages of rainbow trout, *Salmo gairdneri. Aquatic Toxicology and Hazard Evaluation.* 634:231-240.

Klaverkamp, J.F., Hobden, B.R. (1980) Brain acetylcholinesterase inhibition and hepatic activation of acephate and fenitrothion in rainbow trout *Salmo gairdneri. Canadian Journal of Fisheries and Aquatic Sciences.* 37:1450-1453.

Kristofferson, R., Broberg, S., Oskari, A., Pekkarinen, M. (1974) Effect of subletal concentration of phenol on some blood plasma enzyme activities in the pike (*Esox lucius* L.) in brackish water. *Annales Zoologici Fennici.* 11:220-223.

McKim, J.M., Christensen, G.M., Hunt, E.P. (1970) Changes in the blood of brook trout (*Savelinus fontinalis*) after short-term and long-term exposure to copper. *Journal of the Fisheries Research Board of Canada.* 27:1883-1889.

Nemcsøk, J., Benedeczky, I., Boross, L., Asztalos, B., Orbàn L. (1981) Subcellular localization of transaminase enzymes in fishes and their significance in the detection of water pollution. *Acta Biologica Szeged.* 27:9-15.

Nemcsøk, J., Boross, L. (1982) Comparative studies on the sensitivity of different fish species to metal pollution. *Acta biol. Acad. Sci. Hung.* 33:23-27.

Nemcsøk, J., Nemeth, **A.,** Buzò, Zs., Boross, L. (1984) Effect of copper, zinc and paraquat on acetylcholinesterase activity in carp (*Cyprinus carpio* L.). *Aquatic. Toxicology.* 5:23-31.

O'Brien (1960) *Toxic Phosphorus Esters.* Academic Press, New York.

Olàh, J., Janurik, E., Nemcsøk J. (1986) *The Utilization of Thermal Waters for Fish Farming.* Special publication of the Hungarian Fisheries Research Institute (Szarvas).

Phipps, L.G., Holcombe, G.W., Fiandt, J.T. (1981) Acute Toxicity of Phenol and Substituted Phenols to the Fathead Minnow. *Bulletin of Environmental Contamination and Toxicology.* 26:585-593.

Reichenbach-Klinke, H.H. (1972) Histologische und enzymatische Veranderungen nach Schadstoffeinwirkung beim fisch. Veröffentlichungen Institute Kuesten Binnenfischerei-Hamburg. 53:1124.

Reitman, S., Frankel, S. (1957) *Amer. J. Clin. Path.* 28:56.

Rojik, I., Nemcsøk, J., Boross, L. (1983) Morphological and biochemical studies on liver, kidney and gill of fishes, affected by pesticides. *Acta Biol. Acad. Sci. Hung.* 34:81-92. 5.

Peroxidase Activity in Macrophytes as a Marker of Aquatic Pollution

by Sashwati Roy and Osmo Hänninen
Department of Physiology, University of Kuopio, SF 70211, POB 1627, Kuopio, Finland

Purpose

Plants are an integral and the most abundant component of the ecosystem. Their stationary nature makes them preferable models for long-term monitoring programs (Roy and Hänninen, 1993). Moreover, they are quite inexpensive to use. The community composition of macrophytes differs with the gradation from polluted to unpolluted waters; using this criteria macrophytes are extensively used to infer the trophic status of a water reservoir (Kurimo, 1970; Caffrey, 1987). Plants are sensitive indicators of their habitat. Enzymes of plant exposed to pollutants are efficient markers of contaminated environmental conditions (Castillo *et al.*, 1987; Roy and Hänninen, 1992).

Among the conventional biomonitoring methods, biochemical monitoring enables an early detection of adaptive responses in exposed organisms before any structural changes occur, diseases appear or population shifts take place (Roy and Hänninen, 1993).

The main purpose of this method is to rapidly and economically monitor the levels of toxic waste released into the aquatic environment.

Principle

The initial effects of a toxic compound on the organism are normally manifested by changes at the biochemical level of cell function before visible morphological alterations appear. As almost all the biotransformation reactions in cells are catalyzed by enzymes, exposure to xenobiotics often involves disturbances in enzyme function and expression. The consequences of the change (i.e., increase or decrease in activity or concentration) and relative extent of the effect make these parameters a suitable tool for monitoring environmental pollution.

Peroxidases (EC 1.11.1.7), which are ubiquitous in the plant kingdom, are one of the principal enzymes oxidizing xenobiotics in plants with a wide spectrum of substrate specificity (Lamoureux and Frear, 1979). Phenolic compounds are one of the chief constituents of pulp mill effluents (Hutchins, 1979). Phenols are also among the most common substrates for peroxidase which undergo oxidative condensation. Changes in the activity of peroxidase in plants provided valuable information about the state of water reservoirs polluted by pulp and paper mill effluent (Roy and Hänninen, 1992).

Description

Sampling programme

The macrophyte species present in a study area are collected upstream and downstream from the point of discharge of the contaminants. The species are sampled from at least 8-10 locations of each selected site within the study area. Only those plants which visibly appeared to be healthy are collected.

Treatment of material

All samples are preliminarily processed at the site of collection. Leaf and root (if available) tissue from the collected plant species is selected as the experimental material. Material thus collected is first washed with lake water and subsequently with distilled water to remove mud and other adhering detrital matter. The specimens are then frozen in liquid nitrogen and subsequently stored at -80°C on returning to the laboratory on the same day.

Determination of peroxidase activity

Frozen plant tissues are homogenized in 0.2 M Tris-HCl buffer, pH 7.0, containing 7.5% polyvinylpyrrolodine (PVP) w/v. The homogenate is passed through four layers of cheese cloth and the filtrate centrifuged for 15 min at 20,000 g. The supernatant thus derived is used as a source for the assay of peroxidase activity. The processing is done at 0°C to +4°C. Peroxidase activity is assayed spectrophotometrically in a mixture containing the enzyme source, 3.4 mM guaiacol as substrate, 0.9 mM H_2O_2 and 50 mM sodium phosphate buffer (pH 6.0). The rate of increase in absorbance is measured at 470 nm with a double beam spectrophotometer. The activity is then calculated using an extinction coefficient of 26.6/mM/cm at 470 nm for tetraguaiacol (Putter, 1975). The protein content of the enzyme source is determined according to Lowry *et al.* (1951) using crystalline bovine serum albumin as a standard.

Specimen Results

The activity of peroxidase has been used as an indicator of pollution stress in the fine roots of *Pinus sylvestris* and in the macerated leaf tissue of forest trees (Keller, 1984; Castillo, 1986; Markkola *et al.*, 1990). The extracellular activity of this enzyme has also been suggested to be a reliable indicator of environmental stress (Castillo *et al.*, 1987). We have also found that activity of the enzyme is strongly related to the pollution (caused by pulp and paper mill effluent) tolerance capabilities of aquatic macrophytes (Roy *et al.*, 1992). Elevation in the activity of the enzyme was found in the leaves and roots of *Eichhornia crassipes* exposed to pulp mill effluents (Roy and Hänninen, 1992) and also in plants exposed to 2 µM pentachlorophenol (original data submitted for publication by Roy and Hänninen, 1993).

Application

The method can be used in the monitoring of contaminations released in the aquatic reservoirs. The method has been tested in the aquatic reservoirs polluted with pulp and paper mill effluents (Roy and Hänninen, 1992) and as a bioassay to assess the toxicity of sediment-incorporated metal contaminants present in a water reservoir (Klaine *et al.*, 1993). It is, however, expected to have broader significance. The use of other plant enzymes (responsive to pollutants) may be additional potential indicators of polluted environmental conditions.

Complementary information

The sensitivities and responses vary greatly in different organisms on exposure to pollutants. Thus, a wide spectrum of species in an integrated biochemical program with macrophytes will be more reliable than the use of one species only. Practical incorporation of diverse biochemical responses of various species on exposure to pollutants is required in the biochemical monitoring programs of aquatic environment. Plants are easily available and inexpensive to use, therefore the total cost of such biomonitoring programs is rather low.

References

Caffrey, J.M. (1987) Macrophytes as biological indicators of organic pollution in Irish rivers. pp. 77-87. In: Richardson, D.H.S. (ed.) *Biological Indicators of Pollution.* Royal Irish Academy, Dublin.

Castillo, F.J. (1986) Extracellular peroxidases as markers of stress. In: Greppin, H., Penel, C., Gaspar, T. (eds) *Molecular and Physiological Aspects of Plant Peroxidases*, Centre de Botanique, Univ. of Geneve, Geneve. pp 420-426.

Castillo, F.J., Miller, P.R., Greppin, H. (1987) Extracellular biochemical markers of photochemical oxidant air pollution damage to Norway spruce. *Experientia.* 43:111-115.

Hutchins, F.E. (1979) Toxicity of pulp and paper mill effluent. EPA-600/3-79-013.

Keller, T. (1984) The influence of SO_2 and CO_2 uptake and peroxidase activity. *Eur. J. For. Path.* 14:354-359.

Klaine, S.J., Sutton, H.D., Byl, T.D., Bailey, F.C. (1993) Aquatic macrophyte bioassays: Laboratory, *in-situ* and indigenous plant bioassays for sediment toxicity evaluation. Abstract. In: *Ecotoxicology and Environmental Chemistry - a Global Perspective*, First SETAC Congress, Lisbon, Portugal, March, 1993.

Kurimo, U. (1970) Effect of pollution on the aquatic macroflora of the Varkaus area, Finnish Lake District. *Ann. Bot. Fennici* 7, 213-254.

Lamoureux, G.L., Frear D.S. (1979) Pesticide metabolism in higher plants: *in vitro* enzyme studies. pp. 77-128. Paulson, G.D., Frear, D.S., Marks, E.P. (eds) In: *Xenobiotic Metabolism: In vitro methods*. vol.97, American Chemical Society Symposium Series. Washington, D.C.

Lowry, O.H., Rosebrough, N.J., Lewis, Farr, A., Randall, R.J. (1951) Protein measurement with the Folin phenol reagent. *J. Biol. Chem.* 193: 265-275.

Markkola, A.M., Ohtonen, R., Tarvainen, O. (1990) Peroxidase activity as an indicator of pollution stress in the fine roots of Pinus sylvestris. *Water, Air and Soil Pollut.* 52:149-156.

Putter, J. (1975) Peroxidases. pp. 685-690. Bergmeyer, H.U. (ed.) In: *Methods of Enzymatic Analysis.* Verlag Chemie Weinheim Academic Press, Inc. New York and London.

Roy, S., Ihantola, R., Hänninen, O. (1992) Peroxidase activity in lake macrophytes and its relation to pollution tolerance, *Environmental and Experimental Botany*. 32(4):457-464.

Roy, S., Hänninen, O. (1992) Biochemical Monitoring of the Aquatic Environment using Macrophytes, In: *Proceedings of the Symposium on the State of the Environment and Enviromental Monitoring in Northern Fennoscandia and the Kola Peninsula, Rovaniemi*, invited.

Roy, S., Hänninen, O. (1993) Biochemical monitoring of the Aquatic Environment: Possibilities and Limitations, In: *Ecotoxicology Monitoring*. Richardson, M. (ed.) VCH Publishers Ltd. London, U.K. in press.

Mussel Behaviour in the Indication of Water Pollution

by Janos Salanki
Balaton Limnological Research Institute of the Hungarian Academy of Sciences,
H-8237 Tihany, Hungary

Principle

Mussels inhabiting both marine and freshwater ecosystems are permanently in contact with the water and with chemicals dissolved in it. Their life depends on the constituents of the water, and they react to changes of water quality with adequate modulation of their filtering behaviour. By recording shell movements indicative of water pumping activity one can collect information on the quality of the water, especially the presence and effect of some types of environmental pollutants. Shell movements can be quantitatively evaluated.

Morpho-functional Basis of the Filtering Activity

A general feature of mussel behaviour is the rhythmic and periodic character of the shell movements, which is accomplished by the active contraction of the adductor muscles and the oppositely directed passive tension of the ligament connecting the two shell valves. The adductors connect the two valves in such a way that their contraction causes closure of the shell (Fig. 1A). Contrary to this, the elastic ligament connects the valves at one side and the elastic force acts against the adductors. The contraction of the adductor muscles prevails strongly against the force represented by the ligament, however, when the adductor muscles relax, the force of the ligament enables the shells to open easily (Fig. 1B). This means that opening and closing of the shells is regulated actively only through the regulation of muscle contraction and relaxation.

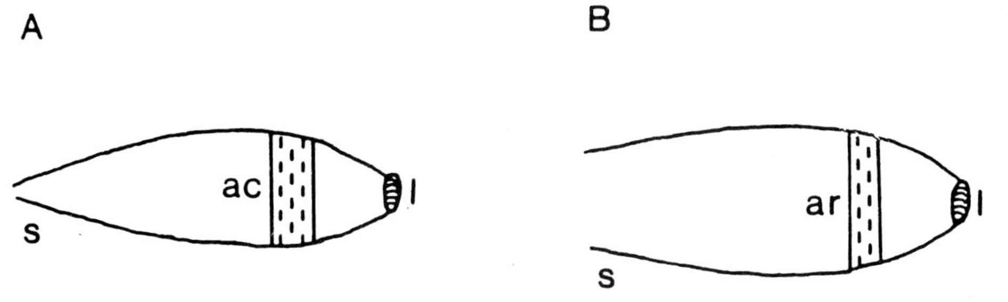

Fig. 1. Schematic representation of the structures involved in the closing (A) and opening (B) of the shell in mussels. ac - adductor muscle contracted; ar - adductor muscle relaxed; l - ligament; s - shell.

Bivalves may have one or two adductors and their location is characteristic to the species. The freshwater mussel, *Anodonta cygnea* L. has two adductors: anterior adductor and posterior adductor. Their movements are coordinated by the central nervous system, which is composed of three pairs of ganglia: the cerebral, visceral and pedal ones. The ganglia are connected by relatively long connectives (Fig. 2). Adductors are innervated from the closest located ganglion, but there are nerve fibres arriving from the more distant centre as well (Salanki and Gubicza, 1969).

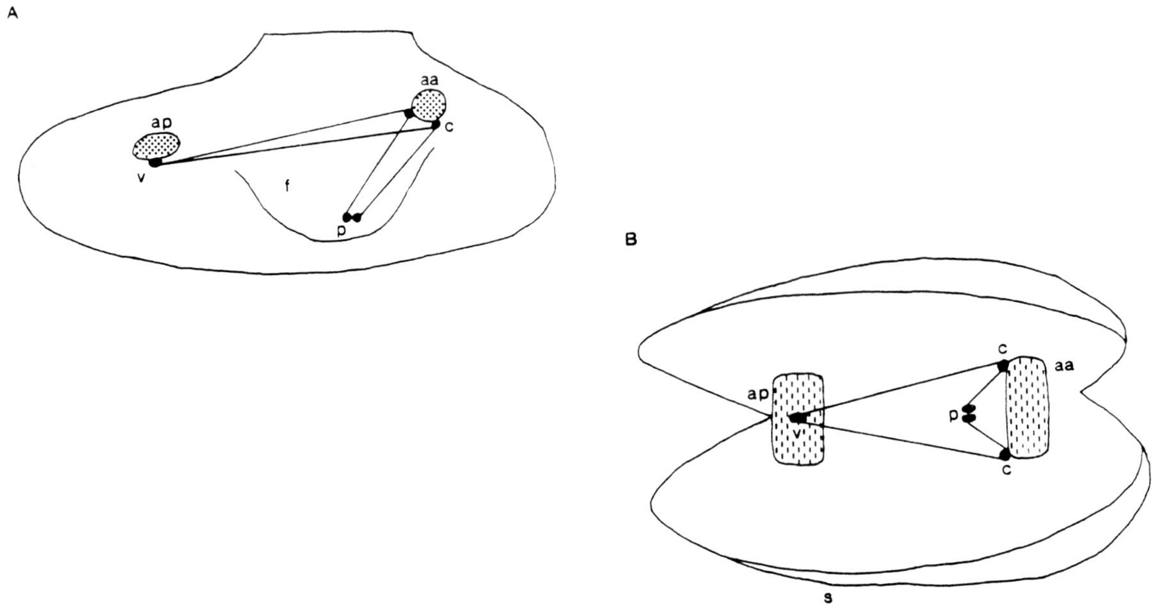

Fig. 2. Schematic representation of the adductor muscles, the central nervous system (ganglia) and the connectives in the mussel *Anodonta cygnea* L. Vertical (A) and horìzontal (B) view. aa - anterior adductor; ap - posterior adductor; c - cerebral ganglia; v - visceral ganglia; p - pedal ganglia; f - foot.

The Rhythmic and Periodic Activity

Mussels are filter-feeders (Bayne,1976; Purchon,1977;) i.e., they take up food from the water which passes through their mantle cavity. Oxygen is taken up from the same water by the gills suspended in this cavity. The opening and closing of the shell valves is a principal component of the filter-feeding behaviour. The open position of the shell allows passage of water containing food particles and oxygen between the gills and towards the opening of the intestinal tract (mouth). During closure of the shell the animal is separated from the outside water. Both the open and closed position of the shell may last for hours and alternates with some regularity. The alternation of open and closed state is called periodic activity. During open period sequences of comparatively fast closing and opening of the shells occur, which represent pumping movements (Fig. 3) Pumping helps the passing and renewal of the water in the shell cavity.

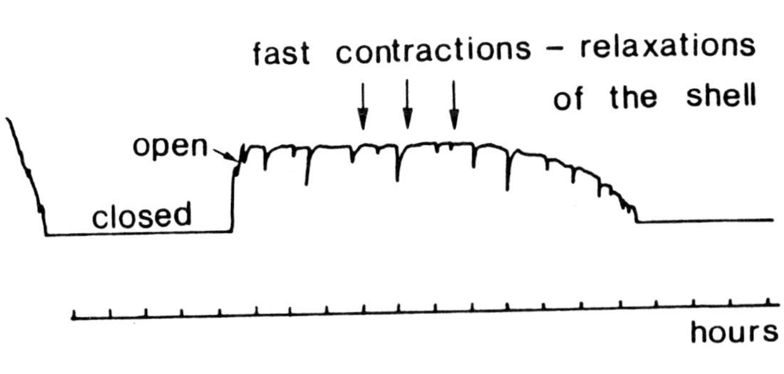

Fig. 3. Record of rhythmic and periodic activity of bivalves.

Recording of mussel activity

The basic requirement is to transmit the relative position and movements of the shell valves to a recording system in a continuous way.

In laboratory conditions the simplest method is when one of the valves is fixed and the the other is connected to a mechanical pen recorder (Fig. 4). The movement of the lever can be traced on a slow (2-5 cm/h) moving recording paper (Salanki and Balla,1964). The animals are kept in water, preferably with a perfusion system, through which various chemicals can be added and constant concentration can be assured.

Recording the shell movements of free moving (unfixed) animals in both laboratory and natural conditions can be performed by electromechanical transmission, based on the inductive attenuator principle. In such a case, a small source and sensor should be mounted to the two valves (Véro and Salanki, 1969).

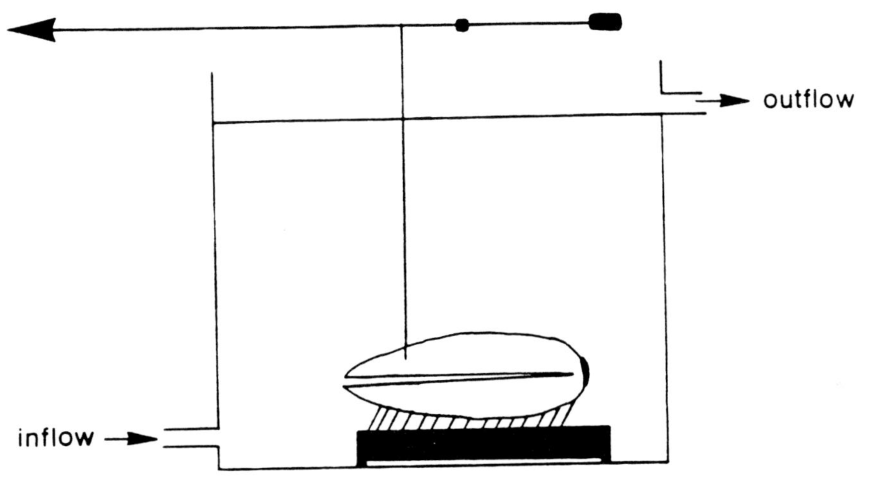

Fig. 4. Principle of recording the shell movements.

Data collection and evaluation

For evaluating the changes in the rhythmic and periodic activity relatively long recordings are necessary (from two days up to several weeks). Using recording paper its speed should not be slower than 2 cm/h, otherwise fast contractions will fuse. On the other hand, in the case of long experiments it is unpractical to use fast-speed recording (more than 5 cm/h).The amplitude of the recording from closure to full opening should be between 2-4 cm.

Evaluating the records one should measure the duration of the open position ("activity") and closed position ("rest") of the shell, as well as the frequency of the rhythmic contractions during activity. The values can be expressed in relation to the day length (24 hours) or in arbitrary units (comparing the duration of active and rest periods, or comparison of periods before and after intervention). See details in next paragraph.

Collecting and evaluation of data can be performed in a more sophisticated way using a computerized system with appropriate programme (Kramer *et al.*, 1989).

computerized system with appropriate programme (Kramer *et al.*, 1989).

Effect of pollutants on mussel behaviour

Since the animal's behaviour (frequency of rhythmic and periodic activity) varies between individuals (Salanki *et al.*, 1970), self-controls should be used for checking the pollutant's effects. The behaviour of the animal has to be recorded at least two days prior to adding the substance to be tested. When the substance or polluted water is introduced into the vessel containing the animal either the duration of active or rest periods, or the frequency of rhythmic contractions may change. As a rule, all three events will change to some degree. Restoration of the original behaviour can take place after eliminating the chemical agent by returning to clean water perfusion. In many cases restoration is not complete or needs several days, due to long effect and accumulation of pollutants.

The behaviour of the animal under the effect of the pollutant should be recorded usually for two-three days for a reliable evaluation.

Examples

1. Data expressed in relation to day length (Fig. 5)
Values to be estimated:
A - duration of active periods (hours)
R - duration of rest periods (hours)
A - average duration of active periods
R - average duration of rest periods
Fc - mean frequency of rhythmic contractions during active periods (cph)
A/24 - portion of active time in a day (%)
R/24 - portion of rest time in a day (%)

2. Data expressed in relation to control (Fig. 6)
Values to be estimated:
A - mean duration of active periods during control time (one week)
R - mean duration of rest periods during control time (one week)
A' - mean duration of active periods after adding pollutants (one week)
R' - mean duration of rest periods after adding pollutants (one week)
A'/A - fractional change of active periods under the effect of pollutants
R'/R - fractional change of rest periods under the effect of pollutants

Conclusions

Recording and evaluation of mussel behaviour has been used in studying the effect of heavy metals successfully, showing that Cd-, Cu-, Hg- ions affect strongly both active and rest periods, while Pb- and Zn- ions are less effective, and influence mainly active periods (Salanki, 1989; 1992). Other chemicals, like insecticides, also cause changes in the animal's behaviour (Salanki and Varanka, 1978) being indicative to the environmentally harmful effects of these substances.

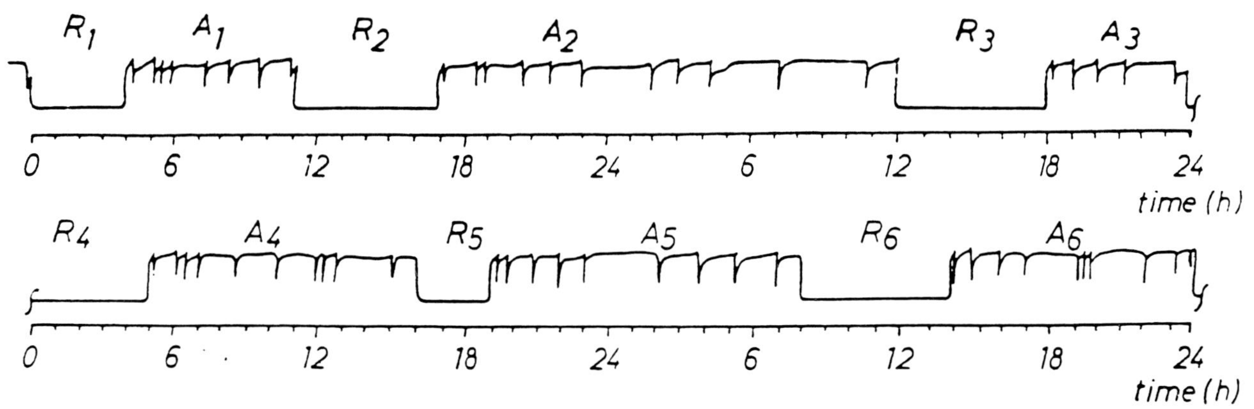

$$\bar{A} = \frac{A_1 + A_2 + A_3 + A_4 + A_5 + A_6}{6} = \frac{7 + 19 + 6 + 11 + 13 + 10}{6} = 11\ \text{h}$$

$$\bar{R} = \frac{R_1 + R_2 + R_3 + R_4 + R_5 + R_6}{6} = \frac{4 + 6 + 6 + 5 + 3 + 6}{6} = 5\ \text{h}$$

$$A/24 = \frac{A_1 + A_2 + A_3 + A_4 + A_5 + A_6}{4 \times 24} \times 100 = \frac{7 + 19 + 6 + 11 + 13 + 10}{4 \times 24} \times 100 = 68{,}75\ \%$$

$$R/24 = \frac{R_1 + R_2 + \dots + R_6}{4 \times 24} \times 100 = \frac{4 + 6 + 6 + 5 + 3 + 6}{4 \times 24} \times 100 = 31{,}25\ \%$$

$$Fc = \frac{\text{sum of fast contractions}}{\text{active time /h/}} = \frac{55}{66} = 0{,}83$$

ADDITION OF POLLUTANT

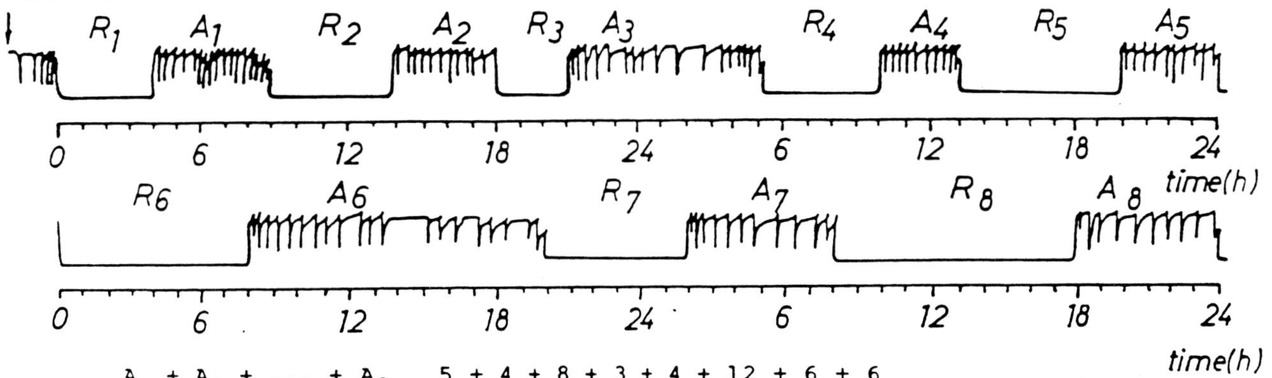

$$\bar{A} = \frac{A_1 + A_2 + \dots + A_8}{8} = \frac{5 + 4 + 8 + 3 + 4 + 12 + 6 + 6}{8} = 6\ \text{h}$$

$$\bar{R} = \frac{R_1 + R_2 + \dots + R_8}{8} = \frac{4 + 5 + 3 + 5 + 7 + 8 + 6 + 10}{8} = 6\ \text{h}$$

$$A/24 = \frac{A_1 + A_2 + \dots + A_8}{4 \times 24} \times 100 = \frac{5 + 4 + 8 + 3 + 4 + 12 + 6 + 6}{4 \times 24} \times 100 = 50\ \%$$

$$R/24 = \frac{R_1 + R_2 + \dots + R_8}{4 \times 24} \times 100 = \frac{4 + 5 + 3 + 5 + 7 + 8 + 6 + 10}{4 \times 24} \times 100 = 50\ \%$$

$$Fc = \frac{108}{48} = 2{,}25$$

Fig. 5. Evaluation of the time relations of periodic and rhythmic activity as a fraction of the day (24 hours). R1-6 - consecutive rest periods; A1-6 - consecutive active periods; fc - fast contractions during active periods. Both in "control" and "after addition of pollutants" four day record is demonstrated and evaluated.

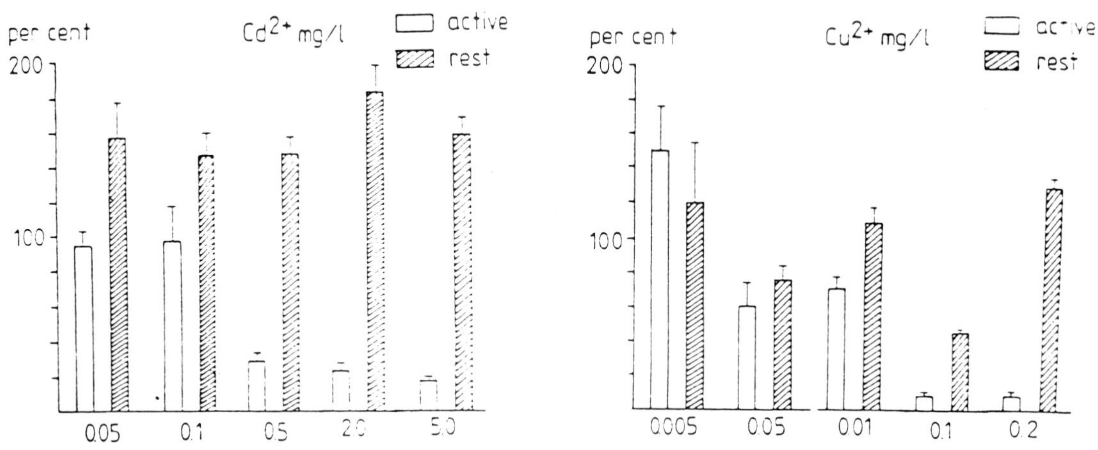

Fig. 6. Evaluation of the effect of pollutants by comparison of the mean duration of active and rest periods before and after introduction of polluted water. The mean values (hours) of one week control recording are taken as 100 per cent both for the active and rest periods. These mean values varied between 11.5±0.81 and 29.0±9.63 hours for the active periods and between 2.84±0.28 and 9.45±0.82 hours for the rest periods in the cases demonstrated. The diagrams show the effect of different Cd^{2+} and Cu^{2+} concentrations. Treatment lasted for one week.

References

Bayne, B.L. (ed.) (1976) *Marine Mussels: Their Ecology and Physiology*. Cambridge University Press. Cambridge.

Kramer, K.J.M., Jenner, H.A., de Zwart, D. (1989) The value increment response of mussels: a tool in biological monitoring. *Hydrobiologia*. 188/189:433-443.

Purchon, R.D. (1977) *The Biology of the Mollusca*. Second edition. Pergamon Press. Oxford.

Salanki, J., (1989) New avenues in the biological indication of environmental pollution. *Acta Biol. Hung*. 40:295-328.

Salanki, J. (1992) Heavy metal induced behaviour modulation in mussels:possible neural correlates. *Acta Biol. Hung*. 43:377-388.

Salanki, J., Balla, L. (1964) Ink-lever equipment for continuous recording of activity in mussels (mussel-actograph). *Annal. Biol.* Tihany. 31:109-116.

Salanki, J., Gubicza, A. (1969) Histochemical evidence of direct neuronal pathways in *Anodonta cygnea* L. *Acta Biol. Acad. Sci. Hung*. 20:219-234.

Salanki, J.,, Glaizner, B., Labos, E. (1970) On the temporal organization of periodic and rhythmic activity in freshwater mussel (*Anodonta cygnea* L.). *J. Interdiscipl. Cycle Res.* 1:123-134.

Salanki, J., Varanka, I. (1978) Effect of some insecticides on the periodic activity of the freshwater mussel (*Anodonta cygnea* L.). *Acta biol.Acad.Sci.hung*. 29:173-180.

Vèro, M., Salanki, J. (1969) Inductive attenuator for continuous registration of rhythmic and periodic activity of mussels in their natural environment. *Med. and Biol. Eng.* 7:235-237.

Methods for Investigating Membrane Effects of Pollutants in Excitable Cells

by Katalin S.-Rozsa
Balaton Limnological Research Institute of the Hungarian Academy of Sciences,
H-8237 Tihany, Hungary

Principle

The basic biological properties of excitable membranes are similar throughout the animal kingdom. Invertebrate nervous systems have the great advantage of being less complicated than those of vertebrates. Among invertebrates the gastropod brain has become a favourite tool of electrophysiological and neuropharmacological investigations because of its large identifiable neurons. Among freshwater Gastropoda *Lymnaea stagnalis* L. is frequently used due to its large and coloured neurons. Among marine and terrestrial gastropods Aplysia and Helix species are respectively most favoured for this type of work.

Neurons of these animals proved to be sensitive to various kinds of environmental pollutants such as heavy metals and pesticides and they can be used in monitoring harmful effects of other types of chemical agents as well at the level of elementary neural processes.

Methods

Microelectrodes are commonly used for recording membrane and action potentials of single neurons. The microelectrode is a glass capillary pulled out to a tip diameter of 0.1-0.5 µm, and filled with an electrolyte, e.g. potassium acetate or chloride. Such electrodes can penetrate inside cells without damaging the membrane. Glass microelectrodes have a high resistance (in range of 5-20 MOhm) and therefore have to be connected to the recording system (oscilloscope, etc.) via an amplifier with a high-impedance input unit.

Basic apparatus necessary for preparing electrodes, insertion into neurons and recording membrane and action potentials of single nerve cells:
- microelectrode puller
- micromanipulator
- control binocular microscope
- D.C. amplifier with high input impedance
- stimulator for current injection
- oscilloscope and pen recorder.

For storage of data tape recorder or computer system are also used (Fig. 1).

Measurements

Resting potential

A recording of membrane potential is obtained if insertion of the microelectrode into the cell results in an immediate negative potential jump and the potential thereafter remains steady. The value of the membrane potential can vary between -40 to -80 mV.

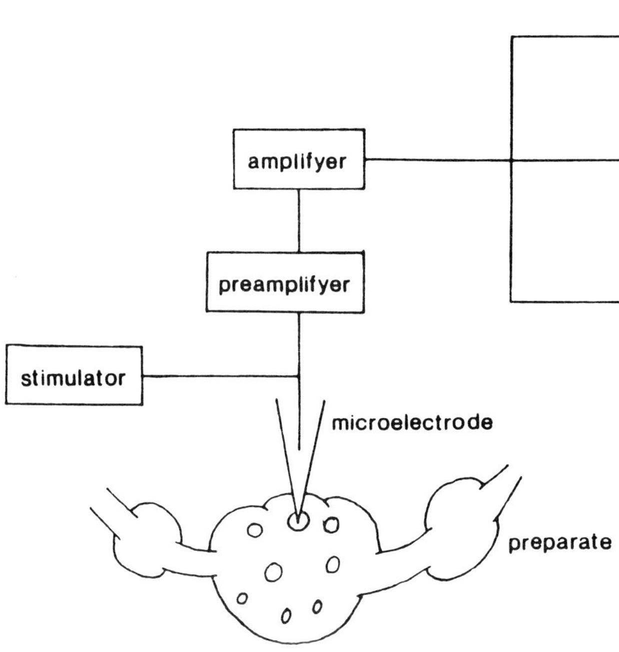

Fig. 1. Block diagram of elementary set-up for recording membrane and action potentials in excitable cells.

Action potential
This reflects the fast changes in membrane potential corresponding to the depolarization and hyperpolarization of the membrane. Action potentials occur when membrane permeability changes to various ions (Na^+,Ca^+,K^+,Cl^-) and an ion current flows in and/or out of the cell. Using a special technique (voltage clamp) direct measurement of ionic current can be performed.

Input resistance
If two microelectrodes are inserted into the same cell, one electrode can be a voltage monitor, while the other can be used to apply current. In the steady state Ohm's law will be applied, i.e., $V=RI$, where V represents the change in membrane potential evoked by the passage of direct current, R is the input resistance and I the intensity of the injected current. R can be determined as V/I.

A detailed account of investigations into the ionic mechanisms underlying resting and action potentials can be found in a number of books (e.g. Kandel, 1976; Levitan and Kaczmarek, 1991).

Practical application

Action of natural chemical substances (transmitters)
Electrophysiological studies on single neurons with chemical agents can either be carried out by including the substance into the bath or by local application from micropipette to

the vicinity of the cell (iontophoresis or pressure injection). Any effect on membrane or action potential and membrane resistance is due either to the direct receptor mediated activation of the neuron from which the recording is made, or due to the release of one or more neurotransmitters from nerve endings (Gerschenfeld, 1973).

Action of environmental pollutants on the voltage dependent currents of the neuronal membrane

Chemical substances occurring in the environment as anthropogenic pollutants may have an effect on the neural membrane and as a consequence on the neural regulation. Their effect can be tested by addition after dissolving in physiological solution to the cell to be studied (Fig. 2). Concentration and time dependency can be determined, and judgement on the harmful effect can be made (S.-Rozsa and Salanki, 1990).

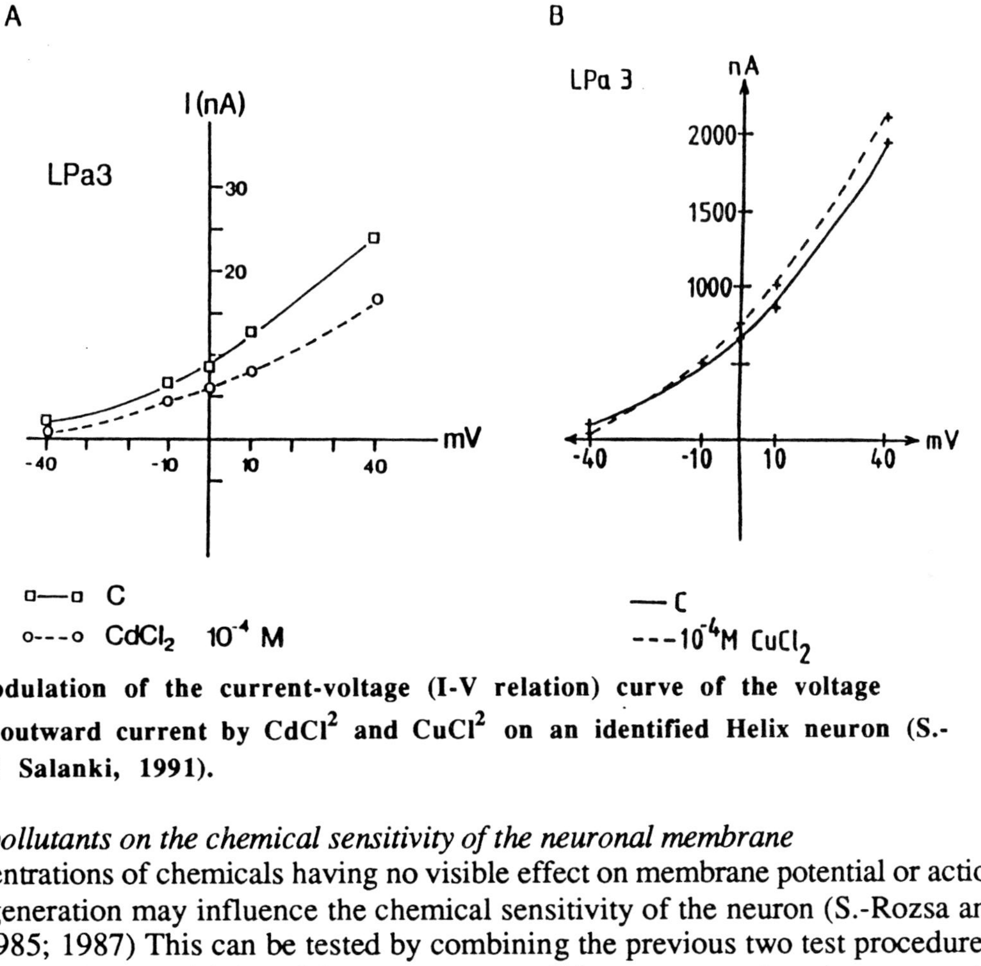

Fig. 2. Modulation of the current-voltage (I-V relation) curve of the voltage dependent outward current by $CdCl^2$ and $CuCl^2$ on an identified Helix neuron (S.-Rozsa and Salanki, 1991).

Action of pollutants on the chemical sensitivity of the neuronal membrane

Low concentrations of chemicals having no visible effect on membrane potential or action potential generation may influence the chemical sensitivity of the neuron (S.-Rozsa and Salanki, 1985; 1987) This can be tested by combining the previous two test procedures: (a) test the chemical sensitivity of the neuron to a well known transmitter (e.g.serotonin, acetylcholin or dopamine); (b) after wash out add the pollutant to be tested; (c) test again the sensitivity to the transmitter with different times (1-20 min) while the pollutant is present; (d) check again the transmitter effect after washing out the pollutant.

Fig. 3 shows an example of heavy metal effect on the chemical sensitivity of central neurons of gastropods.

In a similar procedure the membrane resistance, the ionic currents, the synaptic transmission and other parameters of the membrane can as well be tested.

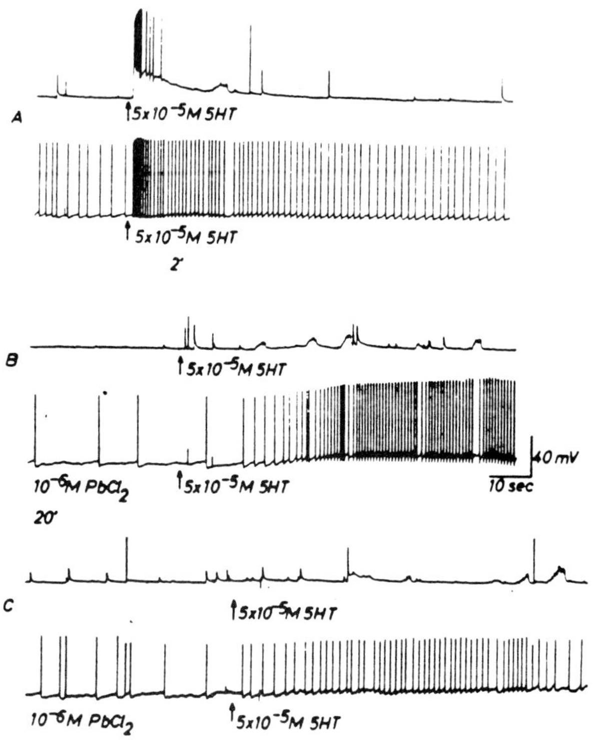

Fig. 3. Effect of lead on the serotonin (5HT) sensitivity of two neurons in the right parietal ganglion of *Lymnaea stagnalis* L.
A - Control activity of the neurons and the effect of 5HT (applied at arrow);
B - Effect of 5HT following 2 min treatment with $PbCl_2$;
C - The same, after 20 min treatment. The 5HT effect is eliminated in the upper cell under the effect of lead, while in the lower neuron only the fast 5HT effect is blocked (S.-Rozsa and Salanki, 1988).

References

Gerschenfeld, H.M. (1973) Chemical transmission in invertebrate nervous systems and neuromuscular junctions. *Physiol.Rev.* 53: 1-119.

Kandel, E.R. (1976) *Cellular Basis of Behaviour. An Introduction to Behavioural Neurobiology.* Freeman Comp., San Francisco, U.S.A.

Levitan, I.B., Kaczmarek, L.K. (1991) *The Neuron. Cell and Molecular Biology.* Oxford University Press, New York, U.S.A.

S.-Rozsa, K., Salanki, J. (1985) Effects of heavy metals on the chemosensitivity of neuronal somata of *Lymnaea stagnalis* L. In: Salànki, J. (ed.) *Heavy Metals in Water Organisms.* Akadémiai Kiado, Budapest. pp 387-400.

S.-Rozsa, K., Salanki, J. (1987) Excitable membranes - objects for evaluating the effect of heavy metal pollution. *Acta Biol. Hung.* 38:31-45.

S.-Rozsa, K., Salanki, J., Présing, M. (1988) Use of *Lymnaea stagnalis* in monitoring heavy metal pollution. In: Yasuno, M., Whitton, B.A. (eds.) *Biological Monitoring of Environmental Pollution.* Tokai University Press, Tokyo, pp 247-255.

S.-Rozsa, K., Salanki, J. (1990) Heavy metals regulate physiological and behavioural events by modulating ion channels in neuronal membranes of molluscs. *Environmental Monitoring and Assessment,* 14: 363-375.

S.-Rozsa, K., Salanki, J. (1991) Ion channels of nerve membrane as targets for environmental pollutants. In: Jeffrey, D.W., Madden, B. (eds.) *Bioindicators and Environmental Management.* Academic Press Limited, London, pp 389-399.

Cytochrome P450 Induction: Ethoxyresorufin O-Deethylase Activity (EROD) in Monitoring Environmental Quality

by John J. Stegeman [1] and Pirjo Lindström-Seppä [2]
(1) Department of Biology, Woods Hole Oceanographic Institution, Woods Hole, MA 02543, U.S.A., and (2) Department of Physiology, University of Kuopio, SF 70211, P.O.B. 1627, Kuopio, Finland

Estimating the degree of ecosystem contamination by toxic organic chemicals can be accomplished through analysis of biochemical changes. One of the most widely used now is the induction of a specific enzyme, that is a particular member of a family of proteins collectively known as cytochrome P450. Numerous studies are showing that the amount of the inducible protein in organisms can be related in a quantitative way to the degree of contamination of the organism by chemical inducers. Here we provide a brief background on the evidence supporting the use of this change as a marker of exposure, the types of chemical contaminants that the marker can indicate, and a guide to the methods for analysis of this biochemical change. The comments here refer particularly to aquatic species and systems, but the approaches are now being applied to other species and systems as well.

Cytochrome P450 (or simply P450) refers to a family of enzymes that catalyze monooxygenase reactions which can transform the structure of organic chemicals (for review, see Ortiz de Montellano, 1986). In eucaryotic organisms the P450 proteins are membrane bound. Those that metabolize or transform xenobiotic chemicals are primarily in the endoplasmic reticulum; upon disruption of the cell, the ER fragments into small vesicles (microsomes). As of 1991, more than 150 P450 genes or proteins from procaryotes and eucaryotes have been sequenced, and organized into 28 gene families (Nebert *et al.*, 1991). P450s are classified as to family, subfamily and member, for example P4501A1 or CYP1A1 is family 1, subfamily A, member 1. By altering chemical structures, cytochrome P450 enzymes may render a given compound nontoxic or, by contrast, drastically increase its toxicity. The amounts of some types of P450 can be induced (increased) in response to an organism's exposure to many types of chemical, including many substrates. As a result, the rate of chemical transformation catalyzed by these enzymes is altered.

P450 induction can also serve as a highly sensitive indicator of an organism's toxic burden, or the extent to which it has been exposed to chemical inducers in the environment. Knowledge of the details of the biochemistry of cytochrome P450 is essential to the interpretations that might be made regarding the induction of P450 as a bioindicator. It is not our intent to provide that detail here. Such detail can be found in numerous other sources. It is rather our intent to give a guide to the literature and a guide to the practical aspects of the analysis. The description and methods that we provide are taken from the primary literature and from several reviews, cited below.

The structure and function of microsomal cytochrome P450 proteins that metabolize foreign compounds have been studied most extensively in mammalian systems, primarily in liver. Fish and invertebrate species possess microsomal P450 systems which are in many ways similar to those present in mammals. The basic biochemistry of the microsomal P450 systems in aquatic species has been described in considerable detail in several earlier reviews (Bend and James, 1978; Stegeman, 1981; Lech *et al.*, 1982; Lee, 1982; Stegeman, 1989).

Cytochrome P450 Induction

One of the most striking features of P450 systems in vertebrates from fish to mammals, is their inducibility by substrates and by compounds structurally related to substrates (Stegeman, 1981; Kleinow *et al.*, 1987). Induction is a process by which a chemical increases the amount of P450; this generally involves synthesis of new messenger RNA and, subsequently, new enzyme protein. Either P450 enzyme activity, protein, or mRNA might be measured to detect induction.

Catalytic (MFO) activity

Studies over the past 15 years have shown induction of P450 enzyme activity (MFO activity) in fish liver (Payne *et al.*, 1987) by polynuclear aromatic hydrocarbons (PAH) or halogenated aromatic hydrocarbons (HAH). PAH and HAH, especially the planar polychlorinated biphenyls, dibenzofurans and dibenzodioxins include many important and potentially hazardous environmental contaminants. Two activities catalyzed by P450, ethoxyresorufin O-deethylase (EROD) activity and aryl hydrocarbon (benzo(a)pyrene) hydroxylase (AHH) activity, are largely specific in their response to these compounds. These activities occur at very low, often undetectable levels in many control or untreated animals, but are highly induced by treatment with the hydrocarbon compounds. Many other monooxygenase activities are not increased by such treatment, substantiating that EROD and AHH are selectively induced.

Antibodies to P450

Studies in several fish species have demonstrated the presence of multiple cytochrome P450 proteins that are characterized by different properties. These include multiple forms purified from rainbow trout, scup, cod, and other species reviewed elsewhere (Stegeman, 1989). One type of P450 purified from liver of several fish species has been identified as the P450 protein primarily induced by PAH and planar chlorinated aromatic hydrocarbons. The three fish proteins of this type most well-studied to date are scup P450E, trout P450LM$_{4b}$, and cod P450c. These three P450s are structurally similar and catalyze the similar reactions.

Polyclonal and monoclonal antibodies have been prepared to the PAH-inducible P450s from several species. Inhibitory antibodies to the fish PAH-inducible P450s confirm that these are the enzymes responsible for the EROD and AHH catalytic activities; i.e., the MFO activities selectively induced by hydrocarbons. These fish P450 enzymes are also responsible for metabolism of several aromatic hydrocarbon carcinogens.

Antibodies to these fish P450s cross-react with proteins specifically induced by PAH or PCB compounds in many vertebrate species. For example, using Western blot analysis, monoclonal antibody (MAb) 1-12-3 to scup P450E has been shown to strongly recognize single proteins induced by PAH or PCB compounds in vertebrates including fish, birds, and mammals (Stegeman, 1989; Goksoyr *et al.*, 1991). This indicates that such antibodies can be used as reagents for detecting induction of P450.

DNA probes

The first cDNA for fish PAH-inducible P450 was cloned from 3-MC-treated rainbow trout liver, and the sequence was determined by Heilman *et al.* (1988). This sequence confirmed that the PAH-inducible P450 from teleosts is a member of the cytochrome gene subfamily now known as the CYP1A subfamily. The PAH-inducible fish proteins are members of the PAH-inducible P450 gene family, the CYP1 family (Heilman *et al.*, 1988; Stegeman, 1989). Based on earlier arguments (Stegeman, 1989), we refer to this PAH-induced type of P450 in fish as fish CYP1A1. However, the sequence of related proteins in most species is unknown, and the designation as CYP1A1 must be tentative. It is more appropriate to refer to the PAH-inducible fish proteins as CYP1 or CYP1A proteins.

Methods

The induction of CYP1A or P4501A in field or experimental samples may be detected by measurement of catalytic activity, or the amount of protein, or the amount of mRNA. There are several methods for measurement of each. The methods for measurement of catalytic activity are the most widely accomplished available at the present time. Therefore we will provide details on the measurement of EROD activity. Although the methods below consider catalytic activity in liver, the use of immunohistochemistry with detection of CYP1A induced in fixed tissue may be particularly useful. This approach can reveal target cells and organs, and use of fixed material can simplify the collection of samples (Stegeman *et al.*, 1991). We also provide some details concerning the proper handling and preparation of samples. Detailed methods for the other types of measurements are referred to elsewhere (Stegeman, 1987; Stegeman *et al*,. 1992, Goksoyr and Förlin, 1992), for those who may be interested in obtaining these. DNA probes are being evaluated for the detection of messenger RNA that is induced in environmentally exposed animals. The measurement of mRNA could eventually be used widely, but analysis of catalytic and/or immunoassay may at present be more desirable for groups not routinely working with RNA analyses.

Species and site selection

In general, the mechanism of induction of CYP1A appears to be largely similar among vertebrates. Evidence from invertebrates argues that a similar response does not occur in aquatic invertebrates. While most fish species show a strong response to experimental treatment with inducers, there may be some species which do not respond as expected. Use of more than one species is therefore suggested for most situations. The species to be selected will depend, however, on the objectives for the monitoring program. Likewise, the specific sites that might be sampled will depend on the objectives. It may be that selection of two species, and at least two sites are desirable. The two sites should include at minimum one reference and one suspect or contaminated site. Readers are urged to see more detailed discussions regarding the selection of species and sites (Peakall and Shugart, 1993). Once selected, the following should be recorded:

length, weight and sex of animals
weight of liver
weight of gonads

These variables are important as reproductive cycles and condition can strongly affect the response to inducers, particularly in females.

Caging of fish

The use of caged fish in studies of this type has proven successful (Lindström-Seppä and Oikari, 1989; 1990). Caged fish of a similar genetic, dietary and exposure history can reduce much of the variation that can occur in samples from wild populations. The position in the water column and the duration of caging should be considered.

Tissue preparation

The reliability of the various methods will all rely on the suitable preparation and characterization of tissue. Liver of vertebrates has the greatest concentration of the P450 catalyst of the EROD activity, and is easily obtained and prepared. The preparation requires removal of gall bladder, homogenization of the tissue and usually the preparation of subcellular fractions. Other tissues may also be useful, but liver is most commonly used. It is important to note that some portion of the tissue to be analysed and other tissue of the fish (skeletal muscle) should be retained frozen under proper conditions for eventual analysis of chemical residues, if indicated. The protein content of the enzyme preparation must be measured. It is important to use a uniform procedure for the homogenization and centrifugation. Factors to be controlled include:

composition of buffers for homogenation and resuspension
volume of buffer per gram of tissue
homogenizer
rate or method of tissue disruption
method of protein assay.

It is possible to use the complete homogenate as the enzyme preparation. However, most often the activity is greater and more precisely measured if one uses subcellular fractions. These are post-mitochondrial supernatant (PMS) or so-called S-9 preparation, or a microsomal fraction. The PMS fraction is the supernatant of a 9,000 or 10,000 x g centrifugation. The microsomal fraction is the pellet obtained by further centrifugation of a PMS at 100,000 x g. The results obtained with various fractions are comparable (e.g., Munkittrick *et al.*, 1993), but the investigator should be aware that the rates per mg protein will differ depending on which fraction is used. Microsomal fractions should be resuspended in 50 mM Tris-Cl, pH 7.4, 1.0 mM dithiothreitol, 1 mM EDTA and 20% glycerol.

Ethoxyresorufin O-deethylase assay

Ethoxyresorufin O-deethylase or EROD can be analysed by several methods. These are based on the difference in either the fluorescent or the absorbance characteristics of the substrate 7-ethoxyresorufin and the product resorufin. The fluorometric and the spectrophotometric assays that are based on these characteristics were originally described by Burke and Mayer (1984), Pohl and Fouts (1980) and Klotz *et al.* (1984). The methods can be employed as continuous assays, using a continuous recording fluorometer or a recording spectrophotometer. However, it is also possible to stop the reactions and measure the amount of product formation at a single time point. The continuous measurements will reveal the kinetics of the reaction, while the stopped reactions may not. Using stopped reactions assumes that the rate of product formation is linear over the time of the assay. This assumption would need to be verified for most studies with novel species. These methods have been adapted to use with a variety of detection devices, including microtitre plate readers. The methods all have common requirements of a suitable enzyme preparation, suitable incubation conditions and a suitable measuring device.

Interpretation and interfering factors

There are a number of biological and environmental factors that can influence the induction of CYP1A, or its catalytic activity (EROD). These are listed as a guide for interpreting the results of activity measurements. The studies that have revealed the influence of these factors can be found in more detailed reviews of the primary literature (Stegeman *et al.*, 1992; others).

Factors that can affect activity of the induced protein:

Reaction conditions:
- incubation temperature
- protein concentration
- substrate depletion

Optimal conditions may need to be determined.

Enzyme preparation:
- enzyme degradation

During tissue preparation or storage. If possible, tissue should be fresh. Otherwise, it should be frozen in liquid nitrogen and stored at - 80°C if possible. Storage in dry ice could be substituted, but there can be loss of activity under these conditions. This may be species specific.
- direct inhibition

Chemicals such as tributyltin or selected PCB congeners that may be present in the liver have been found to inhibit EROD activitiy. This may be more severe in highly contaminated sites. Suspicion of this might recommend the further analysis by immunoassay which is less affected by these compounds.

Factors that are known to affect the induction of CYP1A:
- habitat temperature
- sex and spawning status
- diet

Depending on the questions that are being asked in the monitoring study, there are several courses of action that might be recommended. In the absence of other information, an elevated EROD activity can be considered to reflect exposure to significant concentrations of inducers. Examples of highly potent inducers are:

2,3,7,8-tetrachlorodibenzodioxin
2,3,7,8-tetrachlorodibenzofuran
3,3',4,4'-tetrachlorobiphenyl
benzo(a)pyrene

These and similar structures are considered to be highly toxic and carcinogenic, whether acting by direct or indirect (epigenetic) mechanisms. Detection of elevated EROD activity, P450 protein content or P450 mRNA should at least prompt the consideration of chemical analyses to evaluate the presence of candidate inducers. It should also prompt analysis of other possible effects such as DNA damage.

Analytical Methods

Common Reagents for EROD

7-ethoxyresorufin - There are now several sources, but purity should be checked.

Prepare a saturated solution in methanol, which will have a concentration near 400 μM. A final concentration of 2 μM is used in the assay.

Assay buffer - 0.1 M Tris-HCl, pH 8.0, 0.1 M NaCl (Do not store the enzyme in this buffer, as it will inactivate the P450 after some hours).

NADPH - 25 mg/ml in microsomal resuspension buffer. Store at -20°C or make fresh daily.

Spectrophotometric assay:

Klotz, A.V., Stegeman, J.J. and Walsh, C. (1984) An alternative to 7-ethoxyresorufin O-deethylase activity assay: A continuous visible spectrophotometric method for measurement of cytochrome P-450 monooxygenase activity. *Anal. Biochem.*, 140, 138-145.

Fluorometric kinetic assay:

Burke, M.D. and Mayer, R.T. (1984) Ethoxyresorufin: direct fluorometric assay of a microsomal O-dealkylation which is preferentially inducible by 3-methylcholanthrene. *Drug Metab. Dispos.* 2, 583-588.

Fluorometric end-point assay:

Pohl, R.J. and Fouts, J.R. (1980) A rapid method for assaying the metabolism of 7-ethoxyresorufin by microsomal subcellular fractions. *Anal. Biochem.* 107, 150-155.

Immunoassays (blotting and ELISA):

Goksor, A., Larsen, H.E. and Husoey, A.-M. (1991). Application of a cytochrome P-450 IA1-ELISA in environmental monitoring and toxicological testing of fish. *Comp. Biochem. Physiol. C* 100, 157-160.

Stegeman, J.J., Teng, F.Y. and Snowberger, E.A. (1987) Induced cytochrome P-450 in winter flounder (*Pseudopleuronectes americanus*) from coastal Massachusetts evaluated by catalytic assay and monoclonal antibody probes. *Can. J. Fish. Aquat. Sci.* 44, 1270-1277.

Immunohistochemical assay:

Smolowitz, R.M., Hahn, M.E. and Stegeman, J.J. (1991) Immunohistochemical localization of cytochrome P-450 IA1 induced by 3,3',4,4'-tetrachlorobiphenyl and by 2,3,7,8-tetrachlorodibenzofuran in liver and extrahepatic tissues of the teleost *Stenotomus chrysops* (scup). *Drug Metab. Dispos.*, 19, 113-123.

References

Bend, J.R., James, M.O. (1978) Xenobiotic metabolism in marine and fresh water species. In *Biochemical and Biophysical Perspectives in Marine Biology.* Malins, D.C., Sargent, J.R. (eds). New York, Academic Press, pp 128-172.

Burke, M.D., Mayer, R.T. (1984) Ethoxyresorufin: direct fluorometric assay of a microsomal O-dealkylation which is preferentially inducible by 3-methylcholanthrene. *Drug Metab. Dispos.* 2:583-588.

Goksoyr, A., Andersson, T., Buhler, D.R., Stegeman, J.J., Williams, D.E., Förlin, L. (1991) Immunochemical cross-reactivity of ß-naphthoflavone-inducible cytochrome P450 (P4501A) in liver microsomes from different fish species and rat. *Fish Physiol. Biochem.* 9:1-13.

Goksoyr, A., Förlin, L. (1992) The cytochrome P-450 system in fish, aquatic toxicology and environmental monitoring. *Aquatic Tox.* 22:287-312.

Goksoyr, A., Larsen, H.E., Husoey, A.-M. (1991) Application of a cytochrome P-450 IA1-ELISA in environmental monitoring and toxicological testing of fish. *Comp. Biochem. Physiol. C.* 100C: 157-160.

Heilman, L.J., Sheen, Y.-Y., Bigelow, S.W., Nebert, D.W. (1988) Trout P4501A1: cDNA and deduced protein sequence expression in liver, and evolutionary signicance. *DNA.* 7:379-387.

Kleinow, K., Melancon, M.J., Lech, J.J. (1987) Biotransformation and induction: Implications for Toxicity, bioaccumulation and monitoring of environmental xenobiotics in fish. *Environ. Health Perspect.* 71:105-119.

Klotz, A.V., Stegeman, J.J., Walsh.C. (1984) An alternative to 7-ethoxyresorufin O-deethylase activity assay: A continuous visible spectrophotometric method for measurement of cytochrome P-450 monooxygenase activity. *Anal. Biochem.* 140:138-145.

Lech, J.J., Vodicnik, M.J., Elcombe, C.R. (1982) Induction of monooxygenase activity in fish. In: *Aquatic Toxicology,* Weber, L.J. (ed.). Raven Press, New York, pp. 107-148.

Lee, R.F. (1982) Mixed function oxygenase (MFO) in marine invertebrates. *Mar. Biol. Let.* 2:87-105.

Lindström-Seppä, P., Oikari, A. (1989) Biotransformation and other physiological responses in whitefish caged in a lake receiving pulp and paper mill effluents. *Ecotox. Environ. Saf.* 18:191-203.

Lindström-Seppä, P., Oikari, A. (1990) Biotransformation, other toxicological and physiological responses in rainbow trout (*Salmo gairdneri* Richardson) caged in a lake receiving effluents of pulp and paper industry. *Aquatic Toxicol.* 16:187-204.

Munkittrick, K.R., van den Heuvel, M.R., Metner, D.A., Lockhart, W.L., Stegeman, J.J. (1993) Inter-laboratory comparison and optimization of hepatic microsomal EROD activity in white sucker (*Catostomus commersoni*) exposed to bleached kraft pulp mill effluent. *Environ. Toxicol. Chem.* (in press).

Nebert, D.W., Nelson, D., Coon, M., Estabrook, R., Feyereisen, R., Fujii-Kuriyama, Y., Gonzales, F., Guengerich, F., Gunsalus, I., Johnson, E., Loper, J., Sato, R., Waterman, M., Waxman, D. (1991) The P450 superfamily: Update on new sequences, gene mapping and recommended nomenclature. *DNA Cell Biol.* 10:1-14.

Ortiz de Montellano, P.R. (1986) *Cytochrome P450. Structure, mechanism and biochemistry.* Plenum Press, New York

Payne, J.F., Fancey, L.L., Rahimtula, A.D., Porter, E.L. (1987) Review and perspective on the use of mixed function oxygenase enzymes in biological minitoring. *Comp. Biochem. Physiol.*, 86C:233-245.

Peakall, D.B., Shugart, L.E. (eds). (1993) *Biomarkers: Research and Apllication in the Assessment of Environmental Health.* NATO ASI Series. Berlin, Springer-Verlag, 119 pp.

Pohl, R.J., Fouts, J.R. (1980) A rapid method for assaying the metabolism of 7-ethoxyresorufin by microsomal subcellular fractions. *Anal. Biochem.* 107:150-155.

Smolowitz, R.M., Hahn, M.E., Stegeman, J.J. (1991) Immunohistochemical localization of cytochrome P450IA1 induced by 3,3',4,4'-tetrachlorobiphenyl and by 2,3,7,8-tetrachlorodibenzofuran in liver and extrahepatic tisues of the teleost *Stenotomus chrysops* (scup). *Drug Metab. Dispos.* 19:113-123.

Stegeman, J.J. (1981) Polycyclic hydrocarbons and their metabolism in marine environment. In *Polycyclic Aromatic Hydrocarbons and Cancer*, vol 3, Gelboin, H., TsÛ, P. (eds) Academic Press, New York, pp. 1-60.

Stegeman, J.J. (1987) Monooxygenase systems in marine fish. In: *Pollutant Studies in Marine Animals,* Giam, C.S., Ray, L. (eds), CRC Press, Boca Raton, FL, pp. 65-92.

Stegeman, J.J. (1989) Cytochrome P-450 forms in fish: Catalytic, immunological and sequence similarities. *Xenobiotica.* 19:1093-1110.

Stegeman, J.J., Brouwer, M., Di Giulio R.T., Förlin, L., Fowler, B.A., Sanders, B.M., Van Veld P.A. (1992) Molecular responses to environmental contamination: Enzyme and protein systems as indicators of chemical exposure and effect. In: *Biomarkers. Biochemical, Physiological, and Histological Markers of Anthropogenic Stress.* Huggett, R.J., Kimerle, R.A., Mehrle, P.M. Jr, Bergman, H.L. (eds) Proceedings of the Eighth Pellston Workshop, Keystone, Colorado, July 23-28, 1989. Lewis Publishers, U.S.A., pp. 235-335.

Stegeman, J.J., Smolowitz, R.M., Hahn M.E. (1991) Immunohistochemical localization of environmentally induced cytochrome P4501A1 in multiple organs of the marine teleost *Stenotomus chrysops* (scup). *Toxicol. Appl. Pharmacol.*, 110, 486-504.11.

Hydrobiology

A Simple Biological Index Procedure for Describing and Comparing Estuarine Quality

by D. Jeffrey
Department of Botany, Trinity College, Dublin 2, Republic of Ireland

Purpose

To permit the intercomparison of estuaries as a whole, specific parts of large estuaries and the same estuary over long time periods. The method is specifically designed as a tool for estuarine managers in the context of quality management. It was originally designed for use throughout north western Europe.

Principle

Communities of organisms in estuaries may be classified into either stable communities with relatively long life spans or opportunist communities with short life spans and rapid growth rates. A third designation, "abiotic", applies to areas where no microscopic organisms are apparent. Such areas are frequently either anaerobic or affected by toxic pollution. In the intertidal and benthic parts of an estuary, the broad community types usually occur in mappable zones. Thus the total area of the estuary may be classified and the proportions of each community type are used as the basis for a single number index - the Biological Quality Index (B.Q.I.). This robust index has the great advantage of insensitivity to biogeographic distribution patterns of flora and fauna or even the number and diversity of species. This permits its use over a wide geographic area. It also allows the classification of parts of an estuary with different salinities on a common basis.

Method

1. All available information for the estuary in question should be assembled. This will include geomorphology; flora; fauna; salinity zonation; freshwater loading; nutrient and carbon loading; actual and potential pollutants and contaminants. Appraisal of this information will give a sense of proportion to the survey as a whole and will provide a basis for its logistics.
2. A biological survey of intertidal areas designed to determine the boundaries of the community types is then carried out. With a large system, a training programme for operators is essential. Teams of at least two operators are desirable. They should be provided with field copies of maps or charts of the survey area and a checklist of species. Records should be made of the presence or absence of species, and where appropriate, estimates of organism densities. Attention should be paid to defining boundary or transitional zones between community types.
3. Data from the survey should be interpreted and compiled on a single base map so that the areas occupied by each community type may be defined.
4. From the compilation of data on areas, a Biological Quality Index may be calculated as follows:

Let A represent the 'abiotic' proportion of area; B represent the area occupied by opportunistic communities and C stable communities. Thus A + B + C = 1.0. Biological Quality Index = Antilog10(C-A). The index scale has a maximum value of 10, equivalent to 100% stable communities, down to 0.1 or 100% abiotic.

Examples

Many estuaries in Ireland and some elsewhere in Europe have been surveyed once using this technique as a guide to quality status. In one case a resurvey after a 10 year period has been accomplished in the Tolka basin of Dublin Bay. The Tolka is a small river flowing into a portion of the bay between an industrial and a suburban shore line. It is thus a sensitive part of the whole system. Table 1 demonstrates the nature of change over this period.

Table 1. Changes in Biological Quality Index in the Tolka Basin 1979-1989 (Jeffrey *et al.*, 1991)

		1979	1989	Change
C -	Stable	10%	Zero	-10%
B -	Opportunistic	76%	68%	- 8%
A -	Abiotic	14%	32%	+18%
Biological Quality Index		0.93	0.47	

This shows that the quality of a rather mediocre estuarine area has continued to decline. An objective of water quality management should be to increase the index rating of this specific part of the system. Resurvey of the whole system has been formally included as part of the Water Quality Management Plan for Dublin Bay.

Further notes

The B.Q.I. is only one of a family of estuarine quality evaluation tools. Specific contaminant problems may require a particular biomonitor. We have also developed a complementary non-biological system for indexing the quality of the sediment. This Pollution Load Index may provide specific evidence as to why the B.Q.I. is low or high (Jeffrey and Wilson, 1985).

References

Jeffrey, D.W., Wilson, J.G. (1985) *A Manual for the Evaluation of Estuarine Quality.* 2nd Edition. National Board for Science and Technology, Dublin.

Jeffrey, D.W., Madden, B., Rafferty, B., Dwyer, R., Wilson, J.G. (1991) Indicator organisms as a guide to estuarine management. In: *Bioindicators and Environmental Management*, Jeffrey, D.W., Madden, B. (eds). Academic Press, London.

A Rapid Behavioural Bioassay Based upon the Disruption of Precopula in *Gammarus pulex* (L.)

by D. Pascoe and M.J. Poulton
School of Pure and Applied Biology, University of Wales, College of Cardiff,
P.O. Box 915, Cardiff CF1 3TL, U.K.

Purpose of Procedure

Although field and laboratory bioassays of water quality have conventionally relied upon lethal responses of test organisms, it is clear that the use of sublethal criteria could facilitate detection of lower pollutant concentrations. The disruption of precopula in *Gammarus pulex* (L.) is a behavioural response which has been shown to occur rapidly at low toxicant concentrations and in a concentration related manner. The technique can be used in the laboratory for predicting the toxicity of new compounds and in the field for monitoring the effects of effluents and other pollutants present in water.

Principle of the Method

G. pulex undergoes a characteristic phase of reproduction where the male and female swim together in pairs for up to ten days, before the female moults and mating occurs. This precopulatory guarding stage is fragile and can be disturbed by environmental stressors, such as a toxicant, to bring about disruption of pairing. Disruption occurs in three stages: (a) Vigorous swimming activity of the male and female for up to two minutes; (b) The male holds the female away from his body by extending his gnathopods, then draws her back into the former position on his ventral surface. This activity is repeated every 20-30 seconds; (c) Male and female "kick" simultaneously by flexing their bodies until the female is released. The time taken for pairs to separate has been found to be directly related to the concentration of toxicants present in the water.

Description of the Method

Two forms of the bioassay have been developed. In the first (DIRECT SEPARATION) the time taken for precopula pairs to separate as a direct result of exposure to the toxicant is recorded. In the second method (INDUCED SEPARATION) pairs are exposed for a fixed time, i.e., before the animals separate in the toxicant and are then placed in a separating medium, 2-phenoxyethanol. The induced separation time, i.e., the time from introduction to the separating medium until both gammarids are seen separately is recorded. The induced separation test is a more sensitive bioassay and is used for lower pollutant concentrations.

Direct separation

1. Obtain precopula pairs of *G. pulex* from the field or from laboratory culture. 20 precopula pairs are required for each control and test solution and for each field site, to

give statistical confidence in the results.
2. Expose precopula pairs to a range of toxicant or effluent concentrations and dilution water as a control.
3. Record toxicant-induced disruption of precopula, i.e., the time from introduction of each pair to the toxicant or effluent until both gammarids are seen separately.
4. Calculate the Median Separation Time (ST50) by time-response analysis (Litchfield, 1949).

Following separation, gammarids can be placed in control water and the time for recovery determined as precopulatory pairing is re-established. This will indicate if the toxicant-induced stress is reversible.

Induced separation

1. Obtain precopula pairs of *G.pulex* as in 1. above.
2. Expose precopula pairs to a range of toxicant or effluent concentrations and dilution water as a control.
3. After a fixed exposure time, i.e. before the animals separate in the test solution (pre-determined in preliminary small-scale tests) transfer containers to the separating medium, 2-phenoxyethanol (2ml l-1).
4. Record the Induced Separation Time, i.e. the time from introduction of each pair to the separating medium until both gammarids are seen separately.
5. Calculate the Median Induced Separation Time (IST50) by time-response analysis (Litchfield, 1949).

Table 1. Median Separation Times (ST50) in minutes, with 95% confidence intervals for *G. pulex* exposed to cadmium.

Recorded Cadmium Concentration (mg/l)	ST50 (min.)	C.I.
<0.001, control	-	-
0.11	1228	371-4061
0.37	470	166-1332
1.21	87	39-194
4.00	23	12-44
12.76	9	5-15

Table 2. Median Induced Separation Time (IST50) in seconds, with 95% confidence intervals for *G.pulex* exposed to cadmium.

Recorded Cadmium Concentration (µg/l)	IST50 (secs)	C.I.
<0.01, control	776	665-881
8.3	798	697-915
10.4	680	580-796
22.0	663	535-820
27.6	518	422-637
36.0	456	374-555
46.7	294	246-350

SPECIMEN RESULTS (from Poulton, Pascoe and Sibun (in prep.) and Poulton and Pascoe (1990). Tables 1 and 2 show Median Separation Times (ST50) and Induced Separation Times (IST50) obtained from the Direct and Induced Separation tests respectively in a laboratory bioassay of cadmium toxicity.

Applications

These methods have been used: (a) in the laboratory for evaluating the toxicity of a range of chemicals; (b) in contaminated pond and stream mesocosms; (c) in natural streams subject to artificially induced pollution pulses; (d) in streams exposed to pollutant discharges.

References

Litchfield, J.T. (1949) A method for rapid graphic solution of time-percent effect curves. *Journal of Pharmacology and Experimental Therapeutics* 97:399-408.

Poulton, M.J., Pascoe, D. (1990) Disruption of precopula in *Gammarus pulex* (L.) - Development of a behavioural bioassay for evaluating pollutant and parasite induced stress. *Chemosphere* 20:403-415.

A Chronic Toxicity Test Using Growth of the Insect *Chironomus riparius* (Meigen)

by E.J. Taylor, S.J. Maund and D. Pascoe
School of Pure and Applied Biology, University of Wales, College of Cardiff,
P.O. Box 915, Cardiff CF1 3TL, U.K.

Purpose of Procedure

Assessment of the toxicity of chemicals to aquatic species has largely relied on evaluating the concentrations which cause acute and/or lethal responses. Standard methods are required to determine the effects on test organisms of chronic exposure to sub-lethal concentrations of toxicants. The growth of larvae of *Chironomus riparius* (Meigen) has been shown to be reduced in a concentration-related manner by low levels of various chemicals. The bioassay can be used in the laboratory to provide information on the toxicity of chemicals and when modified for field use a reduction in growth will indicate the presence of environmental stressors.

Principle of the Method

C. riparius is easily cultured under standard laboratory conditions and has a relatively short life cycle (25 days at 20°C), making it a suitable species for use in routine toxicity testing. The life cycle of *C. riparius* includes four larval instar stages. During the 10-day test period (one-third of the life cycle), 2nd instar larvae maintained under control conditions develop to the 4th instar stage. An accurate evaluation of the chronic effects of toxicants can be made by comparing the growth of these control animals to that of larvae exposed to toxicants.

Description of Method

1. Obtain 10 egg masses from a laboratory culture within 12 hours of oviposition.
2. Incubate eggs as a single culture at 20°C for 10 days, after which period the culture contains approximately 2000 2nd instar larvae.
3. Randomly allocate individuals to compartments (6 cm^3) of repli dishes containing 4 ml of solution. 50 larvae are required for each test concentration and control.
4. Place cellulose-mulch substrate, previously soaked in the relevant test solution, in the compartments to a depth of 3 mm.
5. Each day renew solutions and feed each larva with 0.5 mg of finely ground synthetic fish food (Tetramin-TetraWerke Germany) administered as 0.5 ml of a 1g/l suspension in test solution.
6. Conduct the 10-day test in an incubator with a photoperiod of 16 hours light at 20°C.
7. At the end of the exposure period (control larvae will have developed to 4th instar) blot larvae dry and measure the final wet weights of individuals.
8. Analyse results by ANOVA and Unplanned Multiple Comparisons to determine the no observed effect concentration (NOEC) and lowest observed effect concentration (LOEC) of the toxicant.

Specimen Results (from Taylor *et al.*, 1991)

Figure 1 illustrates the effects of a range of concentrations of the insecticide lindane on growth of *C. riparius*.

Table 1 shows the NOEC's and LOEC's of four reference chemicals determined from the growth bioassay performed in the laboratory.

Applications

The bioassay of *C. riparius* growth has been used in the laboratory to evaluate the chronic toxicity of a range of chemicals. Investigations have also been performed in a stream mesocosm contaminated with 3,4-dichloroaniline and field trials with polluted streams are underway.

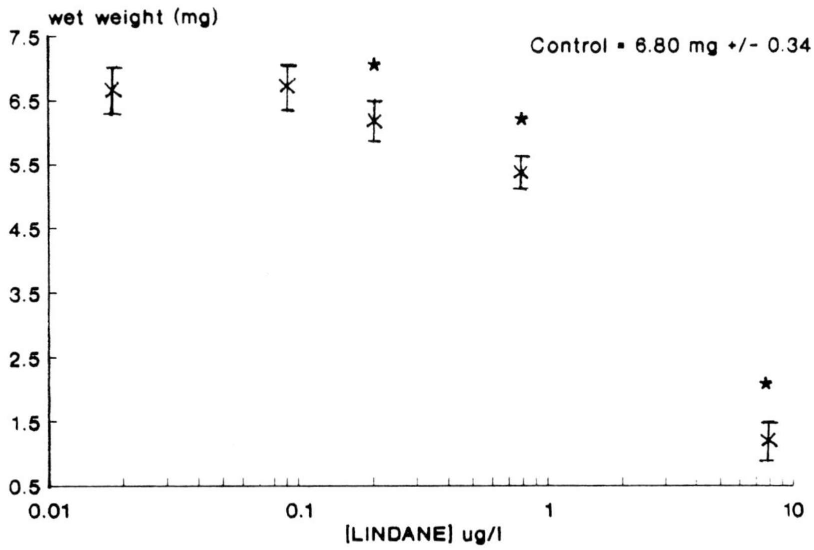

Fig. 1. Mean Wet Weight with 95% confidence limits for *C. riparius* after a 10-day exposure to lindane. * indicates a significant difference (P = 0.05) from the control.

Table 1. Lowest (LOEC) and no observed effect concentration (NOEC) for 10-day growth of *C. riparius* exposed to DCA, atrazine, copper and lindane.

	DCA	ATRAZINE	COPPER	LINDANE
	(mg/l)	(mg/l)	(μg/l)	(μg/l)
LOEC (growth)	0.97	4.9	25.3	0.20
NOEC (growth)	0.17	2.0	16.9	0.09

Reference

Taylor, E.J., Maund, S.J., Pascoe, D. (1991) Evaluation of a chronic toxicity test using growth of the insect *Chironomus riparius* Meigen. *Proc. Sixth International Bioindicators Symposium*, Dublin 1990.

Monitoring of Heavy Metal Pollution with Freshwater Invertebrates

by M. Yasuno and S. Hatakeyama
National Institute for Environmental Studies, Onogawa, Tsukuba 305, Japan

Introduction

Organisms which accumulate pollutants are useful for biomonitoring of the state of environments. It has been emphasized that they are superior to the chemical analysis of water samples, since the concentration of pollutants in water may fluctuate while organisms are continuously exposed to the pollutants. Mussels are useful as a bioaccumulator to monitor heavy metal pollution in coastal areas. It is also desirable to use such organisms in fresh waters. However, most invertebrates are intolerant to heavy metal pollution. Several species of macroinvertebrates have been used for this purpose. Among them, *Baetis thermicus* (Ephemeroptera, Insecta) is an efficient accumulator.

Materials

Baetis thermicus: It is distributed widely in Japan and closely related species can be found in most countries. This species lives in riffles of rivers grazing attached algae. It has a higher tolerance to heavy metal pollution compared with other macroinvertebrates. The species is common in non-polluted rivers also and is collected easily.

Hydropsyche or Chumatopsyche: These Trichopteran larvae are commonly found both in heavy metal polluted rivers and non-polluted rivers. Species of these genera can be collected without difficulty in most continents or regions. Therefore, they are the second candidate as a monitoring invertebrate.

Methodology

Collection

Sampling of macroinvertebrates for this purpose can be made with any type of nets. Approximately 0.5 mm opening mesh nets will be suitable to collect *Baetis* nymphs. Samples are transferred into a white plastic or enamel tray and full grown *Baetis* nymphs are picked up by a pipette. They are rinsed in a small stainless steel cage with 0.1 mM EDTA dissolved in distilled or Milli-Q water to remove metals attached on the body surface. Approximately 30 individuals are pooled in an acid-washed glass vial of 20 ml with forceps. The glass vials, weighed beforehand and numbered individually, should be carried to the field sites. At least three pooled samples are required for each site.

If one wishes to analyse the metal concentration of other macroinvertebrates, samples should be treated as above. As for a large-sized benthos, one or two individuals are sufficient for the analysis. Usually a minimum 10 mg dry weight will be needed to analyse a standard sample, however, it will depend on the analytical limit of equipment.

Laboratory work

Samples in the pre-weighed glass vials are dried at 100°C for 48 hours in an electric oven and then the vials are weighed again to obtain the dry weight of the sample after cooling in a container for desiccation. These samples in vials can be kept in the container for a long period. To digest samples, equal amounts (e.g. 6 ml) of HNO_3 of analytical grade for heavy metals is poured into each of the glass vials containing samples and empty vials for a blank test. These vials are placed on an electric heating plate at 150°C until becoming clear, probably for several hours. If it does not clear, for example, in the case of a large-sized Trichoptera larva, some more HNO_3 is added to the vials and the same amount to the blank test vials. These processes should be conducted in a draft. The digested samples are diluted with distilled or Milli-Q water to 10 to 15 ml and are shaken carefully to provide a homogeneous solution. The exact volume of the solution for each sample is determined from the gravity of the solution and the total weight of solution in the vial. The sample solution (10 to 15 ml) is usually sufficient for triplicate analyses. It is most important that one sample is placed in a single glass vial; they are never transferred to another vial until the end of analysis in the present method. The metal content analysis can be made with this procedure no matter how the samples are injured or decayed during transportation.

Evaluation

The metal concentrations of macroinvertebrates in the highly polluted regions are naturally high and usually proportional to the concentration in the water. However, *Baetis thermicus* is an exceptionally efficient accumulator of cadmium and even when cadmium cannot be detected from water, it will be found in the insect (Fig. 1).

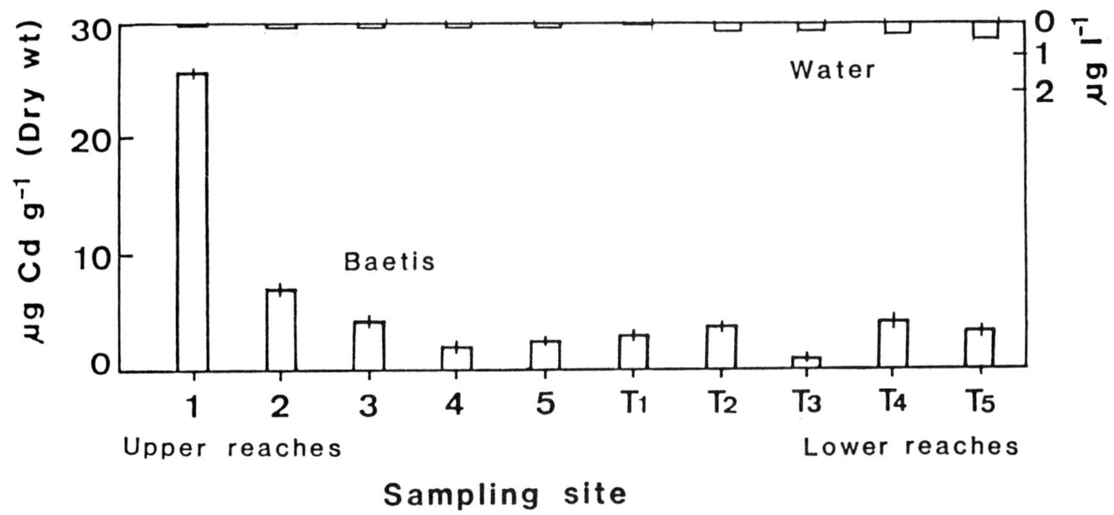

Fig. 1. Cadmium concentrations in water samples and *Baetis thermicus* collected from a river. 1-5: main stream, T1-T5: tributaries.

Biomagnifications through food-chains can be expected in some pollutants, however, the accumulation level of heavy metals by carnivorous macrozoobenthos such as *Protohermes grandis* or Plecoptera is not high. It has been reported that the habitat and feeding habits of invertebrates are rather important in determining the accumulation level of a heavy metal.

Limitation

As the whole body of *Baetis* or other macroinvertebrates is used for the analysis, the gut contents are included. Since *Baetis* is a grazer of attached algae, the heavy metal contents of algea in the gut are not separate.

The possibility of a large-scale downstream drift of macroinvertebrates is low, however, *Baetis* nymphs may drift when they are active or the density is too high. If the copper concentration in water is high, it appears to suppress the accumulation of other metals by macroinvertebrates.

References

Hatakeyama, S., Yasuno, M., Satake, K. N. (1988) *Baetis* (Ephemeroptera) as an efficient heavy metal accumulator in the rivers receiving effluents from abandoned mines. In: Yasuno, M., Whitton, B.A. (eds) *Biological Monitoring of Environmental Pollution*. Tokai University Press. Tokyo, Japan. pp. 237-246.

Smock, L.A. (1983) The influence of feeding habits on whole-body metal concentrations in aquatic insects. *Freshwater Biol.* 13:301-311.

Use of Chironomids as an Indicator of the Environmental State

by M. Yasuno and T. Iwakuma
National Institute for Environmental Studies, Onogawa, Tsukuba 305, Japan

Introduction

Chironomids are commonly found in various environments. They have been used for lake typology for years. Some species may live in extreme environments where no other invertebrates are found. Chironomids are often found in a large number, for example, in the rivers receiving sewage and therefore the appearance of red coloured larvae has been used as an indication of organic pollution.

Materials

Larvae of chironomids are used. Pupal exuviae are also useful and keys for identification to genus are available now. Adult specimens are needed to identify species.

Methodology

Collection and sorting

Equipment

An Ekman-Birge grab is commonly used to collect chironomid larvae in lakes. If larvae inhabit deeper layers of sediment, some device to collect them is necessary as the conventional Ekman-Birge grab will take only the surface sediment. Samples are washed through a wire mesh sieve of 0.06-0.3 mm aperture or a mesh cloth bag of 0.2-0.3 mm to separate larvae.

In rivers, a Surber net or similar is used to collect macroinvertebrates. Since the mesh of the Surber net is not fine enough to collect small size chironomid species, a fine mesh netting sampler is required. Box type samplers are recommended to obtain a quantitative sample. The lateral sides of the box can be made with a transparent acrylic plate, metal plate or wood and the front is open or fitted with a relatively coarse netting to allow the water flow and a conical fine mesh netting is fitted on the back to collect chironomid larvae. To collect small sized chironomid larvae on pebbles or gravel, a 15 cm x 15 cm box sampler with a net of 0.1 mm aperture is used. For medium or large sized larvae, a 25 cm x 25 cm box sampler with a net of 0.2 mm aperture is recommended.

Sampling sites and sampling frequency

In lakes, samplings should be normally made in littoral zones as the species composition is usually different from the profundal one and not comparable. The comparison of the species and their abundance in littoral zones of various lakes will provide valuable information on their richness and diversity.

In rivers, it is preferable that standard samples be taken from the central region of riffles. Chironomids, for the most part, make their nests on either pebbles or rock surfaces, but occasionally some are found living in sand or mud, either with nests, in which case are tubular, or without.

It is desirable to repeat samplings in different seasons to avoid the risk that the sampling is made at the post-emergence period when both the density and standing crops are lowest.

Sample size

More than four samples should be taken at each sampling site in order to calculate the mean and its variance. Two to four extra samples are required for identification to genus or species and enumeration of respective species.

Sample treatment

Samples from lakes are preserved in 5% formalin or in an ice box for further examination and weighing in laboratory. The weighing of fresh or preserved samples should be made after removing water on body surface with filter paper. The fresh weight can be converted into dry weight by multiplying by 0.2. The weight loss or gain during preservation is negligible. Alcohol preservation is not suitable for this purpose.

Two to four extra samples are required to identify genus or species and to enumerate respective species. For this purpose, washed samples can be preserved in a 70% ethylalcohol or isopropanol. Since the biomass of chironomid larvae in rivers will not be compared, samples from rivers can be preserved in alcohol.

Preparation of specimen for classification

Killing larvae or pupae in clean water by heating to near boiling was recommended by Pinder (in Wiederholm (ed.), 1983; 1986) to prepare relaxed mouthparts. Alcohol causes the mouthparts to contract. The mouth part is most important in the taxonomy of chironomid larvae.

Specimens need to be suitably mounted on microscope slides. Larvae are preferably macerated in a hot 10% KOH solution for 5-10 minutes to clear the whole body. The head capsule must be laid horizontally with its ventral side uppermost. As chironomid larvae tend to lie laterally, the head capsule has to be removed by a needle. For long-term storage, they should be mounted in either Canada balsam or Euparal. Otherwise, specimens can be mounted directly by Gum-chloral. In this case, the cover slide must be sealed with a manicure liquid to prevent drying.

Evaluation

Because of the difficulties of species identification, relatively simple methods are described. If one can identify species, further development of the biological indicator may be possible.

Standing crop

This method is recommended for the assessment or classification of the trophic state of lakes. Eutrophication, in general, enhances the primary production, which increases sedimentation and eventually provides more food for benthos while the decomposition of the organic matter decreases the dissolved oxygen at the bottom of lake. As a result, only a few species of chironomids which are tolerant to low oxygen levels breed abundantly. Consequently, the standing crop as fresh or dry weight per unit area (preferably mg/m^2) may indicate the trophic state of the lake (Fig. 1).

Number of species

Number of species of chironomids decreases in accordance with the increase in the pollution level whatever the pollutant (Fig. 2). At the same time, the number of individuals of chironomids tends to increase in polluted regions and therefore the diversity index is usually applicable. Such increase in the individual number is the advantage of using chironomids instead of other invertebrates. Of course, even chironomids cannot live in an extremely polluted region.

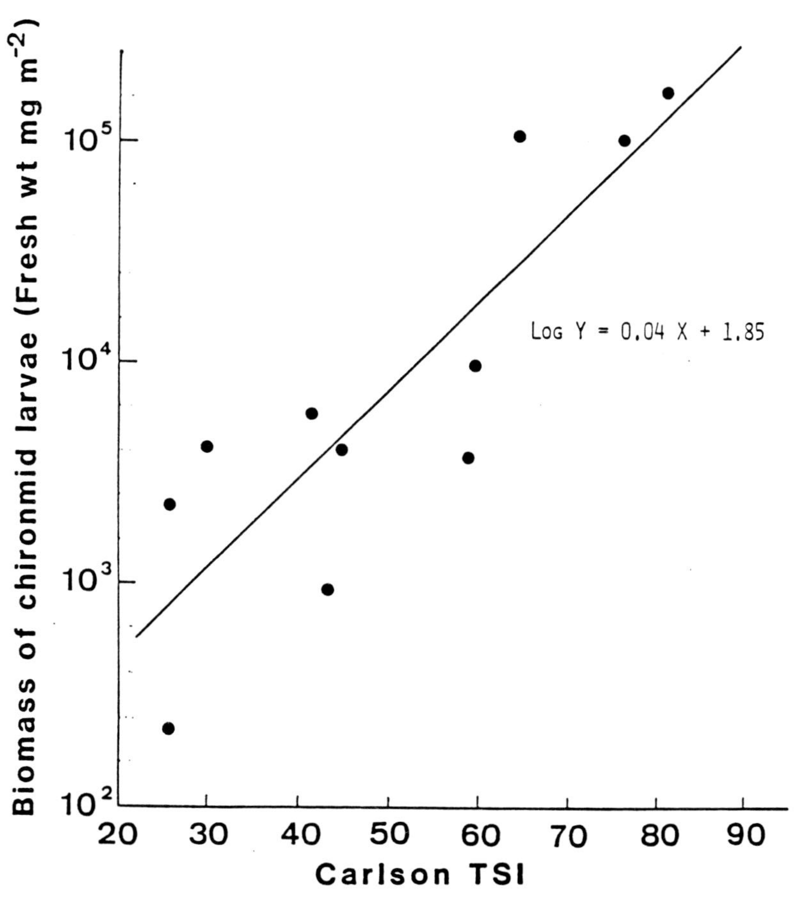

Fig. 1. A relationship between Carlson Trophic State Index and biomass of chironomid larvae in lakes.

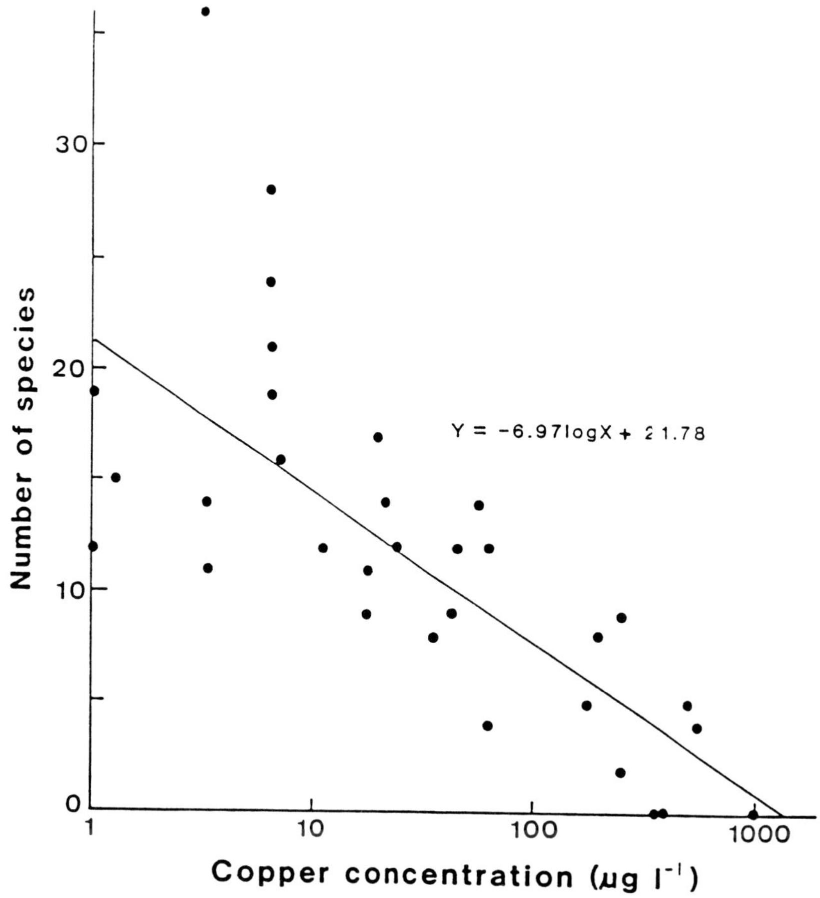

Fig. 2. A relationship between copper concentrations and number of chironomid species.

The species in the rivers polluted with organic matters are usually specific and common in a continent or a region such as *Chironomus thummi* in Europe or *Chironomus yoshimatsui* in Japan. Therefore one can develop the indicator system further, once such specific species are identified.

Limitations

Identification to species level is extremely difficult or impossible, particularly in areas other than Europe or U.S.A. However, it is possible to distinguish the different species with one careful observation. A list of species specific to pollution in one continent or region may not be applicable to others.

Seasonal variation in the abundance must be considered for the method based on the standing crop.

Reference

Wiederholm, T. (ed.) (1983) (1986) (1989) Chironomidae of the Holarctic region. Keys and diagnoses. Part 1. Larvae, Part 2. Pupae, Part 3. Adult. *Entomologica Scandinavica Supplement* Nos. 19, 28, 34.

Index